The Politics of Shale Gas in Eastern Europe

Fracking is a novel but contested energy technology – so what makes some countries embrace it while others reject it? This book argues that the reason for policy divergence lies in procedures and processes, stakeholder inclusion and whether a strong narrative underpins governmental policies. Based on a large set of primary data gathered in Poland, Bulgaria and Romania, it explores shale gas policies in Central Eastern Europe (a region strongly dependent on Russian gas imports) to unveil the importance of policy regimes for creating a 'social licence' for fracking. Its findings suggest that technology transfer does not happen in a vacuum, but is subject to close mutual interaction with political, economic and social forces; and that national energy policy is a matter not of 'objective' policy imperatives, such as Russian import dependence, but of complex domestic dynamics pertaining to institutional procedures and processes, and winners and losers.

ANDREAS GOLDTHAU is Professor in International Relations and Director of the Centre of International Public Policy at Royal Holloway, University of London.

Cambridge Studies in Comparative Public Policy

The **Cambridge Studies in Comparative Public Policy** series was established to promote and disseminate comparative research in public policy. The objective of the series is to advance the understanding of public policies through the publication of the results of comparative research into the nature, dynamics and contexts of major policy challenges and responses to them. Works in the series will draw critical insights that enhance policy learning and are generalizable beyond specific policy contexts, sectors and time periods. Such works will also compare the development and application of public policy theory across institutional and cultural settings and examine how policy ideas, institutions and practices shape policies and their outcomes. Manuscripts comparing public policies in two or more cases as well as theoretically informed critical case studies which test more general theories are encouraged. Studies comparing policy development over time are also welcomed.

The Politics of Shale Gas in Eastern Europe

Energy Security, Contested Technologies and the Social Licence to Frack

ANDREAS GOLDTHAU
Royal Holloway, University of London

CAMBRIDGE
UNIVERSITY PRESS

CAMBRIDGE
UNIVERSITY PRESS

University Printing House, Cambridge CB2 8BS, United Kingdom

One Liberty Plaza, 20th Floor, New York, NY 10006, USA

477 Williamstown Road, Port Melbourne, VIC 3207, Australia

314–321, 3rd Floor, Plot 3, Splendor Forum, Jasola District Centre, New Delhi - 110025, India

79 Anson Road, #06-04/06, Singapore 079906

Cambridge University Press is part of the University of Cambridge.

It furthers the University's mission by disseminating knowledge in the pursuit of education, learning, and research at the highest international levels of excellence.

www.cambridge.org
Information on this title: www.cambridge.org/9781107183940
DOI: 10.1017/9781316875018

First published 2018

Printed in the United Kingdom by Clays, St Ives plc

A catalogue record for this publication is available from the British Library

ISBN 978-1-107-18394-0 Hardback

Contents

List of Figures — *page* viii
List of Tables — ix
Acknowledgements — x
List of Abbreviations — xi

1 Introduction: Shale Gas, Energy Security and Comparative Public Policy — 1
- 1.1 The Eastern European Shale Gas Puzzle — 1
- 1.2 The Comparative Public Policy of Shale Gas — 4
- 1.3 Why Study Shale Gas (in Eastern Europe)? — 8
- 1.4 Research Strategy and Empirical Data — 12
- 1.5 Structure of the Book — 15

2 The Policy Context: European Energy Security and Russian Import Dependence — 18
- 2.1 Europe's Shale Envy: The Triple Premium of America's Energy Revolution — 18
- 2.2 Energy Security in Central Eastern Europe: Russia and Gas Import Dependence — 22
- 2.3 Shale in the Context of European Gas Market Dynamics — 31
- 2.4 EU Energy Policy-Making: Between Multilevel Governance and National Prerogatives — 38
- 2.5 The Road to Shale: An Eye on the 'Above Ground Factors' — 43

3 The Analytical Context: Policy Regimes and the Social Licence — 48
- 3.1 Regimes in Policy Analysis: A Brief Review of the Concept — 48
- 3.2 Interests, Institutions and Ideas: Defining the Analytical Framework — 51
- 3.3 Operationalizing Policy Regimes: Regime Strength and Shale Gas Policy — 59
- 3.4 The 'Social Licence to Frack' — 65

4 The Stalling Front Runner: Poland 67
4.1 Polish Shale: Policy Context 67
4.2 Actors and Sector Governance 71
4.3 Poland's Policy Approach: Institutionally Path-Dependent, Inclusive and Unwillingly Experimentalist 74
4.4 Policy Narratives: Security, Economy and Jobs 82
4.5 Assessing Poland's Policy Regime: Comprehensive Power Arrangement, Low Institutional Capacity and High-Valence Narratives 88

5 The Naysayer: Bulgaria 91
5.1 Bulgarian Shale: Policy Context 91
5.2 Actors and Sector Governance 95
5.3 Bulgaria's Policy Approach: Top-Down Process Meets Exclusive Power Arrangement 98
5.4 Policy Narratives: 'Environmental Hazard' and 'Economic Sell-Out' Trump 'Energy Independence' 105
5.5 Assessing Bulgaria's Policy Regime: Lacking Interest Representation, Low Policy Ownership and Divergent Frames 111
5.6 Excursus: Bulgarian Shale and the Russia Factor 113

6 A No with Options: Romania 115
6.1 Triangulating the Eastern European Shale Conundrum: The Case of Romania 115
6.2 Romanian Shale: Policy Context 118
6.3 Actors and Sector Governance 122
6.4 Romania's Policy Approach: Non-Participatory, Inconsistent and Devoid of Institutional Leadership 126
6.5 Policy Narratives: Environmentalism and Anti-Capitalism Trump 'Economic Opportunity' 130
6.6 Assessing Romania's Policy Regime: Weak Procedures, Clashing Frames and Political Volatility 136

7 The Comparative Public Policy of Shale Gas in Eastern Europe 139
7.1 Comparing Policy Regimes 139
7.2 Assessing the Social Licence 146

8 Conclusion: Shale Gas, Technology Transfer and Energy Security 150

Appendix 154

References 159

Index 195

Figures

2.1 US unconventional gas production, bcm per year *page* 19
2.2 Natural gas prices 2004–2013, USD/MMBtu 21
2.3 EU gas balance 34
4.1 Poland's gas balance, bcm 68
5.1 Bulgaria's gas demand, bcm 93
6.1 Romania's gas balance, bcm 117

Tables

1.1 Shale gas policy in Eastern Europe (EU) *page* 12
2.1 Russian import dependence of select EU countries, per cent and bcm, 2011 24
2.2 Shale gas reserve estimates and annual consumption, bcm, select countries 30
3.1 Policy regimes in Eastern European shale gas: hypothesis and indicators 64
4.1 Summative assessment of Polish policy regime in shale gas 89
5.1 Summative assessment of Bulgarian policy regime in shale gas 112
6.1 Summative assessment of Romanian policy regime in shale gas 137
7.1 Summative assessment of Eastern European policy regime in shale gas 145
A.1 List of country-level interviews in Poland, Bulgaria and Romania 154

Acknowledgements

The intellectual journey for this book started back in 2012. I would like to thank Mike LaBelle and Benjamin Sovacool for being sparring partners and co-authors on all things Eastern European shale over these years, and for inspiring discussions on the energy security conundrum. I am indebted to Radu Dudau, Atanas Georgiev and Tim Boersma for their excellent feedback on earlier drafts of the manuscript. This book also profited from invaluable input by colleagues and peers at various conferences over the past few years, including the Energy Policy Seminar at Harvard's Kennedy School, workshops at Central European University and the PIR Research Seminar at Royal Holloway. I would also like to thank Agata Hinc, Radu Dudau, Georgi Jetchev, Plamen Peev and Andrius Sytas for their assistance and support in collecting primary empirical data for this book.

A great thank you goes to Meghan O'Sullivan and the entire Geopolitics of Energy Project at Harvard's Belfer Center for Science and International Affairs, as well as the CEU School of Public Policy, for being fantastic hosts for this research project, and for their continued support. I would specifically like to thank Eva Vas, Noemi Kovacs and Zsuzsu Gabor at CEU for putting their unmatched administrative skills to work in managing the IOF grant underpinning this book, and Harvard's Leah Knowles and Lauren Bloomberg for allowing me to make the most of my time at Belfer.

I gratefully acknowledge that the research leading to these results received funding from the People Programme (Marie Curie Actions) of the European Union's Seventh Framework Programme (FP7/2007–2013) under REA grant agreement no. PIOF–GA–2012–331962.

Abbreviations

ANRE	Romanian Energy Regulatory Authority
bcm	billion cubic metres
BEH	Bulgaria Energy Holding
BFIEC	Bulgarian Federation of the Industrial Energy Consumers
BRUA	Bulgaria–Romania–Hungary–Austria interconnector
CEE	Central Eastern Europe
CIS	Commonwealth of Independent States
CO2	carbon dioxide
COMECON	Council for Mutual Economic Assistance
E&P	exploration and production
EEA	European Economic Area
EIA	Energy Information Agency
ENTSO-G	European Network of Transmission System Operators for Gas
EPA	Environmental Protection Agency
ETS	EU Emissions Trading System
EU	European Union
EWRC	Energy and Water Regulatory Commission (Bulgaria)
FSU	Former Soviet Union
GDOS	General Directorate for Environmental Protection (Poland)
GDP	gross domestic product
GECF	Gas Exporting Countries Forum
GHG	greenhouse gas
GISC	Civil Society Initiative Group (Grupul de Iniţiativă al Societăţii Civile) (Romania)
IBR	Romania–Bulgaria Interconnector
IEA	International Energy Agency
IGB	Interconnector Greece–Bulgaria
IMF	International Monetary Fund

IOC	international oil company
LNG	liquefied natural gas
LTC	long-term contract
MEE	Ministry of Economy and Energy (Bulgaria)
MEW	Ministry of Environment and Water (Bulgaria)
MMBtu	one million British thermal units
MWh	megawatt hours
NAMR	National Agency for Mineral Resources (Romania)
NAPM	National Agency for Environmental Protection (Romania)
NGO	non-governmental organization
NIMBY	not in my backyard
NOC	national oil company
NOKE	National Operator of Energy Minerals (Narodowy Operator Kopalin Energetycznych) (Poland)
OECD	Organisation for Economic Co-operation and Development
OPEC	Organization of Petroleum Exporting Countries
PGI	Polish Geological Institute
PiS	Law and Justice Party (Poland)
PKE	Polish Ecological Club (Poland)
PO	Civic Platform (Poland)
PSD	Social Democratic Party (Romania)
RDOS	Regional Directorate for Environmental Protection (Poland)
RIEW	Regional Inspectorate of Environment and Water (Bulgaria)
R/P ratio	reserves-to-production ratio
SLO	social licence to operate
TANAP	Trans-Anatolian Pipeline
TAP	Trans-Adriatic Pipeline
Tcf	trillion cubic feet
TRC	Texas Railroad Commission
TSO	transmission system operator
TTIP	Transatlantic Trade and Investment Partnership
UOKiK	Office for Competition & Consumer Protection (Poland)
URE	Energy Regulatory Office (Poland)

1 *Introduction*

Shale Gas, Energy Security and Comparative Public Policy

1.1 The Eastern European Shale Gas Puzzle

Dan Yergin, the energy historian, once called hydraulic fracturing – or 'fracking' – 'the most important, and the biggest, energy innovation of this century' (New York Times 2013).[1] At least for the United States, this statement holds quite some value. Starting from low production levels in the early years of the twenty-first century, unconventional gas became 'the new conventional' (Trammel 2015) within less than a decade. Soaring domestic gas production from the Permian, Eagle Ford, Barnett, Marcellus and Haynesville shale 'plays' set the country on a firm path towards 'energy independence'. The USA is set to be a net exporter of natural gas this side of 2020 (EIA 2017a). The country now enjoys a significant economic boon, substantially lower carbon emissions (thanks to gas crowding out much dirtier coal) and a national security premium – all of which gives America an 'energy edge' (Blackwill and O'Sullivan 2014).

Unsurprisingly, shale gas has therefore been described as a 'game changer' (Medlock 2009) in global energy, and hydraulic fracturing as a technology that will 'rock the world' (Jaffe 2010). The American 'shale revolution' (Financial Times 2015b) is seen as a role model for import-dependent nations wishing to improve their supply portfolio through domestically available reserves. Globalizing gas markets, in turn, will experience a boost thanks to additional supply. Shale might therefore also fundamentally change the geopolitics of natural gas. Indeed, once the hydraulic fracturing method had matured in the United States, its country of origin, debates emerged over whether the technology might 'go global'. Canada started to produce shale gas, as

[1] Hydraulic fracturing is an extraction technique in which deep rock formations are fractured through high-pressure injection of a 'fracking fluid' and proppants to release the hydrocarbons therein. Coupled with horizontal drilling, fracking is used to exploit unconventional hydrocarbon reserves.

did China and Argentina. Others could follow: according to a widely cited Energy Information Agency (EIA) study, shale reserves are available across the globe (EIA/ARI 2013).

The US shale gas story was observed particularly carefully in Europe. The EU's lopsided import structure has long been an issue of concern: as a bloc, the EU imports some 37 per cent of its gas from Russia (Eurostat 2016). While there exists a longstanding – and, indeed, mostly frictionless – energy relationship between Europe and Russia, dating back to the Soviet Union and the Cold War, the EU's high import dependence on one supplier became a liability in the context of the Ukrainian–Russian gas crises of 2006 and 2009. In the winter of 2009, a complete Russian cut-off left countries in Eastern and South Eastern Europe without gas supplies and tuned import dependence into a security issue in both the Eastern European region and the EU as a whole. What's more, in the wake of the 2008 financial crisis and the resulting economic slope, European industry would clearly benefit from a stimulus similar to the one the US economy had seen thanks to shale.

And yet, despite China and Argentina aspiring to follow in America's footsteps, fracking technology seems to have problems leaving North America, which so far makes the USA and Canada the world's only significant production centres of shale gas (EIA 2015a, 2017b). Particularly in Europe, the technology finds it hard to take hold. Shale gas policies vary significantly across Europe, with some countries pushing ahead and others rejecting fracking altogether. While England and Poland remain committed to their developing shale gas reserves, France, the Netherlands, Scotland, Germany and Bulgaria have enacted bans or de facto moratoria on unconventional gas production. More strikingly, shale gas policies diverge significantly even in Central Eastern Europe (CEE), a region holding promising reserves. CEE countries are mostly dependent on one single supplier – Russia – and against the backdrop of the region's more recent history, energy policy tends to be cast in hard security terms. Yet, Bulgaria has banned shale gas exploration and production (E&P); Poland remains firmly committed to fostering it despite its drawbacks; and exploration in Romania has stalled as attention has shifted to offshore exploration.

This policy divergence is even more puzzling as all these countries share a common regulatory past in Communism, whose legacies are still visible in national energy governance, and in regulatory regimes more generally. Many Eastern European nations also have a historical track record in oil, gas or coal production, as well as in gold and copper mining. In other words, extractive industries are not new to them, and there exists a regulatory apparatus designed to govern hydrocarbon production. What's more, all countries in the region have benefited from the US Unconventional Gas Technical Engagement Program, a policy initiative established under Hillary Clinton's term as Secretary of State. Initially termed the 'Shale Gas Initiative', its aim was to foster the global diffusion of hydraulic fracturing, both to improve the energy security of US allies and to underpin American technology leadership in unconventional gas (Sakmar 2011). Further, as members of the EU, Eastern European countries are subject to an identical supranational regulatory environment in natural gas. EU environmental legislation and three sets of comprehensive 'Energy Packages' define the broader framework in which national energy policy choices happen. Gas market patterns are very similar across the region, too. Incumbent long-term contracts (LTCs) still tend to form the basis of gas trade with key suppliers such as Russia, and the prevalent oil-price peg – in some instances, even outright state price regulation – is only slowly giving way to more competitive pricing arrangements. Finally, national income in all countries in the region remains below the EU average. Put differently, all Eastern European nations have an incentive to foster domestic industry, jobs and tax income by way of nurturing nascent industries such as unconventional energy.

So, what makes some Eastern European countries embrace fracking and others reject it? Standard explanations for a Western European context can by and large be dismissed in the case of CEE. Green parties, for instance, are hardly part of the Eastern European shale conundrum. In none of the CEE member states were the Greens in government when pertinent decisions on fracking were taken, and because they were represented only in some countries' parliaments, their involvement in the political deliberations on energy and shale gas policy has remained limited. Moreover, while environmental movements exist, they tend to be less organized and less powerful compared to Western European member states (Fagan and Carmin 2011). To be sure, as will be discussed later, environmental concerns do play a prominent role in shaping public debates surrounding shale gas, even in Eastern Europe – including the much-debated impact that fracking fluids could have on groundwater safety and habitat – but this does not necessarily translate into well-organized interests on the national level. Party orientation does not seem to make a difference, either. In Poland, it was Donald Tusk's centrist government that pushed shale; in Bulgaria, it was the conservative Boyko Borisov; and in Romania, it was Viktor Ponta, a Socialist. Despite clear commitment from the political leadership, it was only Poland that ended up passing a pro-shale law, whereas Bulgaria eventually banned fracking and Romania remained in a legal halfway house. In short, the explanation for the Eastern European shale gas puzzle needs to lie elsewhere.

As this book will show, it is the distinct way governments interact with private and social actors, and how these interactions are structured through institutional settings and processes, that makes the difference. An important additional factor is the existence of a convincing policy narrative (and whether it is 'taken up' by non-state actors). All three elements determine the degree to which key stakeholders eventually buy into and support shale gas policies as put forward by their governments. With this, the book argues that the causal factors lie in ideas, interests and institutions, rather than in normative motivations or party politics.

1.2 The Comparative Public Policy of Shale Gas

The CEE shale gas puzzle goes right to the heart of comparative public policy: why, despite similar conditions, do national governments end up taking different policy choices (Engeli and Rothmayr Allison 2014; Gupta 2012; Howlett, Ramesh and Perl 2009; Schmitt 2013)? When answering this question, this book not only takes a deep dive into CEE energy policies, but does so from an analytically distinct angle. More to the point, in order to disentangle the complexities characterizing the comparative public policy of Eastern European shale gas, this study builds on the concept of policy regimes. In the broadest sense, policy regimes consist of a power arrangement, a policy paradigm and an organizational arrangement (Wilson 2000) existing around a given policy issue. The power arrangement refers to political and socioeconomic actors, while the policy paradigm frames the problem, and hence also its potential solutions. The organizational arrangement is about institutions and processes in policy formulation and implementation. With this, policy regimes consist of a 'set of ideas, interests, and institutions that structures governmental activity in a particular issue area' (McGuinn 2006).

To be sure, and as we will lay out in more detail in Chapter 3, policy regimes are conceptually heterogeneous. In order to carve out the main analytical argument, the book will therefore primarily rely on the approach taken by May et al. in some of their pertinent recent works (Jochim and May 2010; May and Jochim 2013; May, Jochim and Sapotichne 2011). When disentangling ideas, interests and institutions as the 'governing arrangements for addressing policy problems' (May and Jochim 2013, 429), focus is placed on investigating the role played by domestic (incumbent) actors, the (regulatory) framework they act in and the ideational drivers guiding national policy discourse and action. More to the point, the book suggests that it is the strength of the policy regime that makes the difference and explains divergent national shale gas policy choices. Regimes that align pivotal actors as part of the power arrangement, that do so under a shared vision (the policy paradigm) and that engage key stakeholders in participatory organizational arrangements are more likely to create 'buy-in' opportunities and lend legitimacy to policy goals and processes. Put differently, Eastern European governments were successful in implementing their shale policy agendas in countries where strong policy regimes were in place. Where policy regimes were weak, by contrast, policy agendas failed.

It is important to note that this analysis is particularly interested in the *output* of the policy process; that is, the legal and regulatory frameworks enabling or blocking shale gas development. It is less interested in its *outcome* (notably, whether an unconventional gas industry is about to scale up, or the volumes of natural gas eventually produced) or its *effect* (that is, how policies as adopted might reshape subsequent policy processes). This is for both conceptual and empirical reasons. Conceptually, the material outcomes of a given policy are by and large a question of design (and non-design) (Howlett 2009; Howlett and Mukherjee 2014). A policy's effect is highly contingent not only on the 'rationality' of the policy process but also on external factors that often are outside governmental control. An example here is the international pricing environment, which, in the energy sector, strongly influences investment decisions, independently from domestic regulatory contexts. As a corollary, the question of whether shale gas policies are 'optimal' in terms of achieving a desired end is also not the focus of this study. 'Feed-forward' policy effects, in turn, may become obvious only after some time, are often indirect and may have unintended effects (Jordan and Matt 2014; Schneider and Ingram 2009). Debates about shale in Europe started less than a decade ago; compare that to the twenty-five years it took the unconventional gas sector in the USA to scale up and mature. Moreover, pertinent policies in Poland, Bulgaria and Romania were adopted early in the 2010s; arguably, this was too short a period to allow consistent policy feedback loops to materialize.

Empirically, shale gas exploration in Eastern Europe has by and large been put on hold since 2015, when Chevron, the US energy major, left Romania, marking the end of an exodus of foreign companies that had flocked into the region about half a decade earlier. In other words, in terms of actual production, there simply is not much going on. Reasons lie in so far disappointing geology, adverse policy environments in reserve-holding countries (as discussed later) and, notably, a depressed international market environment. With gas markets going soft (see Chapter 2), Eastern European shale would arguably face an uphill battle even in a more favourable domestic context: if prices are not right, and costs are too high, exploration does not happen. In turn, different price signals, advances in technology and a learning curve from emerging shale industries closer to home (such as in parts of the UK, going forward) might brighten the prospects of unconventional gas in Eastern Europe once more. This implies that a research focus on the policy outcome would be ill advised (as would be writing off commercial shale production in Poland and elsewhere – even if it eventually comes to fruition at a smaller scale than some governments had hoped for). Production levels are a moving target, and are contingent on many factors, including changing external market environments. What matters, instead, is whether pertinent domestic frameworks are in place, or not – the policy *output*. In short, the study object needs to be the national-level policies defining the conditions under which shale E&P may potentially happen, and the contestation around them.

With this, the comparative public policy of shale gas is about the choice of the regulatory frameworks governing shale, and how they are set in place. These frameworks may be favourable to shale development, for instance by way of putting in place incentives to foster E&P, or at least by levelling the playing field. Poland's 2014 Act on a Special Hydrocarbon Tax, for instance, regulates shale gas licensing and taxation, and exempts exploration from tax payments until 2020. Legal frameworks may also be hostile to shale, such as in Bulgaria, where fracking has been banned by an Act of Parliament since 2012. They may even reflect non-choices, such as in the case of Romania, where no decision was taken on fracking, and an initial ban expired without being renewed. As this book will detail, the choice (or non-choice) of shale gas frameworks is a function of the policy regime surrounding unconventional hydrocarbons and the fracking technology. In other words, it is the specific national setting, defined by actor involvement (interests), policy narrative (ideas) and the underpinning procedures (institutions), which determines whether societies take one choice or the other, or none at all.

As a corollary, investigating shale gas is also about the comparative public policy of creating a 'social licence' for fracking.[2] Originating from the literature on environmental protection, the concept of a social licence to operate (SLO) became prominent in works on mining and the extractive industries. In essence, a social licence 'governs the extent to which a corporation is constrained to meet societal expectations and avoid activities that societies (or influential elements within them) deem unacceptable, whether or not those expectations are embodied in law' (Gunningham, Kagan and Thornton 2004, 307). Prno and Slocombe (2012) suggest that a social licence can be considered to exist when extractive operations are met with approval and broad acceptance within a society. It has therefore been likened to a social contract (Giurco et al. 2014).[3] Hydraulic fracturing is not only an extractive technique, it is also highly contested, due to its potential impact on the environment. It can therefore be argued that fracking requires such a social contract to be operated. More fundamentally, it is the creation of pertinent legal frameworks that warrants societal support. For such legal frameworks, a social licence was evidently generated in some countries – the ones embracing the fracking technology – while

[2] I owe this aspect of my argument to the participants of the World Bank Institute Learning Symposium on 'Governance of Unconventional Gas: Exploring How to Deliver Transparent Benefits in Non-OECD Countries', Washington DC, 2–3 June 2014, organized by Philip Andrews-Speed. See also the resulting special issue of *OGEL* on 'The Governance of Unconventional Gas Development Outside the United States of America' (Andrews-Speed 2014).

[3] Despite an emerging literature on the SLO, the concept remains in its infancy, and it is only recently that it has been applied to shale gas (House 2013; Smith and Richards 2015). The literature so far primarily focuses on company-level activities aimed at meeting social and environmental obligations within communities (Owen and Kemp 2013).

in others it was not. As this book argues, policy regimes are key in this context, as they determine whether a proposed (shale) policy is considered legitimate among key stakeholders, whether processes are trusted and whether the stated policy goal and its implementation are credible (Thomson and Boutilier 2011).

1.3 Why Study Shale Gas (in Eastern Europe)?

Shale gas has attracted significant scholarly attention. Because of the highly politicized nature of natural gas, the bulk of the works on shale tends to centre on national and international security aspects, and the implications of a changing natural gas landscape. Shale gas has been subject to investigations in geopolitics (Blackwill and O'Sullivan 2014; Jong, Auping and Govers 2014; Kaplan 2012; Kim and Blank 2014; Kuhn and Umbach 2011; Umbach 2013), national security (Medlock, Jaffe and Hartley 2011) and geoeconomics (Blackwill and Harris 2016; Bros 2012; Haug 2012). These works stand in the tradition of thinking about energy as a means or end of grand strategy, and add to a large set of literature on the nexus of energy, war and peace (Colgan 2013; Kalicki and Goldwyn 2005; Klare 2001, 2009; Shaffer 2009). Reflecting realist or neo-realist approaches to international politics, they hardly open the backbox of (national) energy policy-making, and treat shale gas as an asset in the global competition for influence and supremacy.

The international security lens also dominates works on Eastern European natural gas. Reflecting the fact that energy security features prominently on policy agendas in the region, analyses tend to focus on Russian gas import dependence and energy security concerns (see also Chapter 2). Natural gas is discussed in the context of Russian imperialism (Baev 2008; Orban 2008), Europe's supply challenge (Aalto 2007; Bilgin 2009; Correlje and van der Linde 2006; Finon and Locatelli 2008; Youngs 2009) and pipeline geopolitics (Bahgat 2003; Johnson and Derrick 2012; Stulberg 2012). Many of these works implicitly or explicitly also explore what could be done to counter Russian dominance in CEE gas markets, e.g. by diversifying supply through indigenous sources such as shale. Few studies go beyond the over-dominant security discourse (for an exception, see Kuzemko et al. 2012). Building on the diversification agenda, works also seek to explore whether the US shale gas story could be replicated in a European context and beyond (Boersma and Johnson 2012; Grafton, Cronshaw and Moore 2017; LaBelle and Goldthau 2014a; Nülle 2015).

Leaving the confines of international security, a growing literature starts to address shale gas as a policy phenomenon. Investigations focus on public perceptions of and attitudes towards shale (Alcorn, Rupp and Graham 2017; Boudet et al. 2014; Kriesky et al. 2013; Wolske, Hoffman and Strickland 2013), including risk perception (Graham, Rupp and Schenk 2015; Schafft, Borlu and Glenna 2013); the public discourse and media coverage (Bomberg 2015; Jaspal and Nerlich 2014; Jaspal, Nerlich and Lemańcyzk 2014; Jaspal, Turner and Nerlich 2014); social representation in unconventional energy development (Evensen, Clarke and Stedman 2014; Upham et al. 2015); and the ethics of shale gas policies (Evensen 2016; de Melo-Martín, Hays and Finkel 2014). Many works also investigate policy frames surrounding shale gas and fracking, both for the USA (Lachapelle, Montpetit and Gauvin 2014) and Europe (Cotton, Rattle and Van Alstine 2014; Goldthau 2016c; Metze 2017; Williams et al. 2015), and from a comparative transatlantic perspective (Bomberg 2017). Unpacking the detailed policy dynamics of shale, scholars address unconventional gas in the context of multilayered governance arrangements, notably US federalism (Arnold and Holahan 2014; Burger 2013; Davis and Hoffer 2012; Lin 2014); investigate the regulatory politics of fracking (Davis 2012, 2014; Spence 2013; Warner and Shapiro 2013) and the role of advocacy coalitions (Weible et al. 2016); examine the role of local communities (Neville and Weinthal 2016; Smith and Ferguson 2013); and explore the management of potential risks (Jacquet 2014; North et al. 2014).

Although this brief review remains far from being comprehensive, several patterns emerge from the existing literature.[4] First, there are only a few book-length treatises on shale gas. Available works tend to address the general-interest audience, offering a broad overview of the chances and pitfalls of the industry (Graves 2012), telling the first-level story of the pioneering wildcat 'frackers' and how their energy innovation turned the energy world upside down (Gold 2014; Zuckerman 2013) or adopting an activist stance against the technology (Bamberger and Oswald 2015). A limited number of studies offers more scholarly analyses. While rich in empirical detail, focus here tends to be placed on

[4] For more comprehensive reviews, see Neville et al. (2017) and Sovacool (2014a).

the state- and community-level impact of shale (Gullion 2015; Wilber 2012).

Second, analysis of fracking and shale to date almost exclusively centres on the US experience. This somewhat lopsided focus obviously is a function of America representing the 'motherland' of fracking, and a vast data set being available after some twenty years of consecutive technology deployment across the country. Europe, by contrast, is strongly underrepresented still. Leaving aside the geopolitics literature on shale, available studies primarily cover the UK and Western Europe. The few exceptions include Van de Graaf et al. (2017), who adopt a comparative perspective on European shale policy, and Reins (2017), who offers an EU-level regulatory perspective on new technologies such as fracking. Eastern Europe is almost entirely overlooked in the literature on shale gas policy, with some select works investigating Poland, Bulgaria and Ukraine (Georgiev 2016; Goldthau and LaBelle 2016; Jaspal, Nerlich and Lemańcyzk 2014; LaBelle 2016, 2017; LaBelle and Goldthau 2014b; Lis and Stankiewicz 2016). Although the region features prominently in debates on energy security and Russian geopolitics, surprisingly few scholars go to the trouble of opening the black box and unpacking national-level shale gas politics.

Finally, the comparative public policy literature on energy seems to have a blind spot when it comes to fossil fuel energy technology. To be sure, public policy research abounds on oil, gas, nuclear and renewables, and it would be beyond the scope of this proposal to give comprehensive credit to the available literature on US or EU energy policy, their impact on the share of fossil or renewable fuels in the energy mix and the politics shaping national energy priorities.[5] Recent volumes exploring the technology–public policy nexus, including Grubler and Wilson's (2013) work on energy technology innovation, Ulli-Beer's (2013) study on the governance of energy technology change and Murphy's (2007) investigation into sustainable technology governance, reveal a strong bias towards low-carbon transition. (A clear exception is Smil's (2010) work on energy transitions, whose focus on fossil fuel technology can be explained by the historical perspective it adopts.) What's more, it is particularly the literature building on the various strands of the regime concept – the central analytical

[5] For an assessment of fifteen years of energy scholarship, see Sovacool (2014b).

approach of this book – that tends to place emphasis on the themes of environment and sustainability. This holds true for investigations on the national level (e.g. Howlett 2001), the EU level (e.g. Hunter and Smith 2006) and the global level (e.g. Oberthür and Stokke 2011). This evident research preference may well be a function of the strong contemporary policy impetus towards global sustainability. Still, it leaves a gap to the extent that a central energy innovation of the past decade is by and large left out of pertinent analyses.

In light of the foregoing, this book adds to three distinct scholarly debates. First, it addresses a genuine comparative public policy debate, to the extent that it investigates policy divergence across countries. It focuses on how a specific confluence of policy narratives (ideas), interests (as sources of support or resistance) and organizational arrangements (institutions) shapes domestic policy processes and determines (varying) policy outcomes. Studying Europe – and particularly CEE – also adds important empirical insights into shale gas policy in the context of a multilevel governance framework: the EU. Moreover, the empirical focus on Europe allows the study of the comparative public policy of shale gas in a democratic context. Indeed, Europe represents one of the few cases beyond the USA where elections may punish political leaders for their policies and the rule of law ensures citizen rights and the inclusion of society in political decisions. While countries such as China and Russia sit on vast unconventional hydrocarbon reserves, the development of these reserves will take place in policy regimes that are hardly comparable to Western settings.

Second, the book speaks to an ongoing scholarly and, indeed, political debate on EU energy security. Against the backdrop of high import dependency ratios in natural gas and intensifying conflict surrounding supply and transit routes, CEE has emerged as the focal point of Russia–EU energy (geo)politics. Many hoped that shale gas would become a game changer in Eastern European–Russian energy relations. Unpacking national energy policy, the book provides for an empirical grounded and policy-focused analysis of shale in CEE, a topic that is typically discussed with a strongly normative undertone and approached from a hard security angle. Shedding light on the domestic politics of shale, the book ties energy geopolitics back to the important domestic policy dynamics unfolding between government, civil society and industry. With this, it highlights the local foundations of international security imperatives.

Third, the book contributes to the literature on technology innovation and transfer, using shale gas as a critical case. While most writings in the technology–policy nexus focus on renewables and low-carbon technologies, this book unpacks domestic policy dynamics pertaining to a contested fossil fuel technology. On the one hand, unconventional gas therefore provides a unique opportunity to explore how distinct actor sets and institutional environments play out for a nascent industry in different national contexts, and how a novel technology such as fracking resonates in an incumbent regulatory and policy environment. On the other, it allows the opportunities and obstacles surrounding the global transfer of a contested technology to be unveiled. The Eastern Europe shale conundrum helps shed light on what it might take to make a novel technology take hold in new destinations beyond the USA, and what may make shale 'go global'. This allows conclusions to be drawn on the socio-economic dimension of technology transfer and how a 'social licence' might be generated for the same.

1.4 Research Strategy and Empirical Data

This books considers three cases in detail: Poland, Bulgaria, and Romania. This selection mirrors the cross-section of shale gas policies and the broader population of cases in the region (see Table 1.1). Poland represents a country supporting the development of unconventional gas, and enacted policies to that end. Regardless of whether the much-disputed Polish 2014 'shale gas law' eventually effectuates commercial gas production in the country, Poland retains a firm pro-shale

Table 1.1 *Shale gas policy in Eastern Europe (EU)*

Country*	Shale gas policy	Regulatory framework
Bulgaria	Preventive	2012 moratorium on hydraulic fracturing
Czech Republic	Preventive	2012 moratorium on shale gas licences
Hungary	Neutral	Law on Mining, amended 2014
Poland	Supportive	2014 Special Hydrocarbon Tax Law
Lithuania	Neutral	Underground Law, amended 2013
Romania	Neutral	2004 Petroleum Law

* Countries for which there exist reserve estimates

policy stance and has sought to put in place pertinent frameworks. With this, Poland in fact represents the front runner in the region, and an outlier to the extent that it remains openly positive of fracking. Having banned fracking in 2012, Bulgaria sits on the other end of the spectrum. Sofia not only put a moratorium on hydraulic fracturing, but even extended it to the usage of fracking in conventional gas production. The government also revoked Chevron's exploration permit, which made the company leave the country. With this, Bulgaria marks the most pronounced case of anti-shale policy in Eastern Europe, although other countries have effectively followed a similar path: the Czech Republic, for instance, put a moratorium on granting licences for shale gas exploration until pertinent legislation was in place, and kept on postponing the latter (Reuters 2012b).

Romania, finally, represents a country that has essentially taken a non-choice, in that it has neither banned nor promoted shale; instead, as will be discussed in more detail in Chapter 6, the government pivoted to promoting offshore gas rather than shale. Chevron leaving the country in 2015 ended a heated domestic policy debate without a fundamental decision being taken on the future of available unconventional energy reserves, or the frameworks for extracting it. With this, Romania represents a broader set of countries in the region that have also taken 'non-choices'. This includes Lithuania – the Baltic nation that sought access to the global market for liquefied natural gas (LNG) and recently brought online its first LNG terminal – whose flirt with shale ended in 2014 when Chevron divested from the country's unconventional energy assets, citing regulatory uncertainty (Reuters 2014a). It also includes Hungary, whose government officially remains 'neutral' to shale gas (Landry 2013): although contestations over fracking remain far less pronounced than elsewhere in the region, Budapest has chosen to rely on piped gas for its energy needs, renegotiating contracts with Gazprom in addition to sourcing from the Baumgarten Hub in Austria, as well as pivoting to nuclear in a disputed deal with Russia's Rosatom (Politico 2016).

This cross-section of cases also represents the three most promising reserve-holding countries in the region. The EIA reports of 2011 and 2013 on shale gas reserves outside the USA, which kick-started the excitement over shale gas prospects in Eastern Europe, specifically mentioned Poland, Romania and Bulgaria. They put Poland's unconventional gas reserves at some estimated 4191 billion cubic metres (bcm), or roughly 252 years of the country's annual consumption in 2015; Romania's at 1444 bcm, or 127 years; and Bulgaria's at 481 bcm, or 185 years (see also Chapter 2). With this, the case selection not only includes countries with a significant divergence on policy output – the independent variable of the study – but coincidently also covers the bulk of the estimated reserve base in the region.

Empirically, this book builds on two years of field research in Eastern Europe. It relies on seventy-six structured *sur place* interviews conducted in the region, including in Poland, Bulgaria, Romania and the Baltics, between 2012 and 2014, in addition to numerous unstructured background interviews.[6] Primary data were generated on all levels of governance, including the federal, regional and local levels. Interviewees comprised senior government officials, policy advisors, academics, local mayors, protest group leaders, representatives from public and private companies, business lobbyists and politicians from the entire democratic spectrum.[7] In addition, this study relies on an extensive analysis of pertinent national energy policy documents, corporate assessments, EU reports and media coverage. To the extent that interviews could only be conducted in the local language, native speakers were involved. The same holds true for policy documents. All interviews were transcribed into English so that they could be processed further, notably with a view to identifying dominant policy narratives (see Chapters 4, 5 and 6).

The primary data come with the caveat that some relevant national actors could not be included in the sample. Representatives of state-owned companies in Bulgaria, for instance, were not ready to speak about shale gas due to the political sensitivity of the topic. To the extent possible, other sources were used to make up for the lack of primary data here. Similar issues were encountered with representatives of American companies operating in the region, who only agreed to speak off record. Related data are used as background information but not attributed. Moreover, some interviews, notably with political representatives, had to be conducted anonymously. Throughout the book, the names of individual interviewees are therefore omitted, but their professional function is revealed by way of their position,

[6] A total of of sixty-four interviews were conducted in the country cases studied in this book.

[7] For a complete list of interviews conducted in the country case studies see the Appendix.

seniority level and organizational affiliation. Despite these limitations, this study relies on a rich set of primary data, complemented by secondary data, presenting deep insights into one of the most politicized policy areas in CEE.

In terms of method, the book essentially follows a Most Similar Systems Design. It contrasts three cases exhibiting highly similar conditions that differ on the outcome variable. As detailed earlier, the selected country cases share a Communist regulatory past, have a historical track record in the extractive industries, were all exposed to the US Unconventional Gas Technical Engagement Program, are subject to an identical supranational regulation, operate in similar gas market environments and share the common goal of catching up with Western European welfare standards. Shale gas policies, however, differ greatly, ranging from supporting to rejecting fracking. To disentangle the causal factors at work, the book operationalizes policy regimes and defines indicators that allow their strength to be measured (see Chapter 3). This enables a structured comparison of the three cases, and allows causal inference to be generated between the nature (and strength) of the domestic shale gas policy regime and the policy output in the shape of the regulatory frameworks set in place in the three countries.

1.5 Structure of the Book

The book is organized in six substantial chapters. The next – Chapter 2 – defines the policy context of European shale. It details the three-pronged policy drivers in this area: the attempt to emulate the US 'shale revolution', which brought America the 'triple premium' of an economic boost, a reduced carbon footprint and a gain in international security; Eastern European energy woes around high import dependencies on Russia, notably against the backdrop of gas cut-offs in 2006 and 2009; and the EU's pro-market agenda in energy, which could be underpinned by additional domestic sources of supply. The chapter also introduces European energy policy-making as a multilevel policy field with shared competences in energy and environment. It shows that it is the Commission's 2014 Recommendations, coupled with (mostly) environmental legislation, that provide the EU-level regulatory framework in shale gas, leaving policy decisions by and large to member states. Moreover, by juxtaposing the European situation to that in the USA, the motherland of fracking, the chapter demonstrates that the 'above ground factors' differ substantially. In addition, Eastern Europe features a post-Communist regulatory legacy consisting of large incumbent corporate stakeholders tied to traditional LTCs, regulatory regimes favouring state-run corporations and an energy system mirroring the public utility model.

Chapter 3 develops the analytical framework. It first provides an overview of the regime concept as used in social science inquiry, and specifically in policy analysis. Next, it introduces policy regimes as composed of interests, institutions and ideas, and argues that they may produce cohesive policy dynamics by bridging actor groups and governance levels. The chapter then poses that a causal variable for the success or failure of a governmental policy agenda is contingent on regime strength. A strong regime offers 'buy-in' opportunities for stakeholders, unites them around an overarching policy narrative and lends legitimacy to policy goals and processes. This, it is argued, generates the 'social licence' needed for shale gas policies. Operationalizing policy regime strength, the chapter details three specific hypotheses guiding the empirical investigation.

Chapter 4 delves into the first case: Poland. The empirical investigation reveals a strong path dependency and institutional inertia in shale gas. At the same time, the chapter suggests that various procedural elements pertaining to information, institutional outreach and community empowerment ensured the buy-in of important stakeholders on national and subnational levels. It also highlights the existence of two strong policy frames pertaining to 'national security' and 'economic opportunity'. These narratives are found to resonate strongly with key stakeholders from society, the business community and the state apparatus, on all governance levels. Moreover, they allowed shale gas to be lifted to a 'national project' in terms of enhancing the country's sovereignty and economic prospects.

The empirical investigation of Bulgaria is the subject of Chapter 5. As the analysis suggests, a combination of poorly designed institutional procedures, low administrative capacity and a top-down process characterize Bulgarian shale gas policy. This, the chapter finds, excluded and even alienated key stakeholders, including not only societal groups, non-governmental organizations (NGOs) and environmental movements, but also industry players such as state-owned corporations that stood to lose out. In addition, the government failed to put forward a convincing narrative in support of its policy goals. Consequently, its framing efforts pertaining to security gains and economic benefits saw a low 'ideational uptake' among important stakeholders within Bulgarian society and the economy, and eventually had to give way to competing policy narratives that united a broad opposition against shale.

Chapter 6 turns to Romania. The Romanian case features frequently shifting shale gas policy agendas, impeding a consistent institutional strategy within the state apparatus. This adds to a generally weak public administration in the country, comparably low institutional capacity and non-participatory and in-transparent administrative processes. Post-Communist legacies are found to be strong in Romania, which, coupled with inadequate legal frameworks, resulted in both distrust among the population and poor policy implementation. Governmental policy frames, such as 'economic opportunity', are identified as having had low traction among stakeholders, leaving room for more dominant narratives surrounding anti-neoliberal and environmentally conservative agendas, flavoured with nationalistic undercurrents.

A comparative assessment of the three case studies is offered in Chapter 7. Linking the empirical findings back to the analytical framework, the chapter argues that the shale gas policy regime in Poland was strong, whereas in Bulgaria and Romania it was weak. Although Poland's approach to shale gas policy was also bound by limited administrative capacity, red tape and a lack of coordination between relevant state authorities, flexible institutions and processes, coupled with a 'high-valence' policy narrative, ensured the buy-in of important stakeholders on national and subnational levels. The Bulgarian and Romanian policy regimes, by contrast, featured low institutional capacity, no stakeholder outreach and 'low-valence' policy narratives. Romania, however, eventually turned to the development of Black Sea reserves, which ensured the persistence of the status quo, and allowed the country to retain a legally rather neutral stance on shale, compared to Bulgaria's fracking ban. Based on these findings, the chapter draws conclusions on the creation of a social licence for a shale.

Chapter 8 offers some concluding thoughts on the comparative public policy of contested technologies, and suggests further avenues of research.

2 *The Policy Context*

European Energy Security and Russian Import Dependence

2.1 Europe's Shale Envy: The Triple Premium of America's Energy Revolution

A book on European shale gas must start in the USA. There, unconventional gas production growth soared at more than 50 per cent annually between 2007 and 2012 (EIA 2017d). By 2015, shale gas had come to add 430 bcm to the USA's gas balance per annum, up from 37 bcm in 2007 (see Figure 2.1). While hydraulic fracturing is still very localized in how it is applied on different 'shale plays', as they are each characterized by their own geology, fracking is now a mature technology and has made its way from George Mitchell's Barnett into the rest of America. Shale gas is now produced from plays all over the country, from Texas (Permian, Eagle Ford, or Barnett) to Arkansas and Louisiana (Haynesville), to Pennsylvania, Ohio and West Virginia (Marcellus or Utica play).

The US shale gas revolution put the country on a path towards 'import independence'. Having reduced its imports significantly over the past years, the US Department of Energy's EIA expects the country to become a net exporter of natural gas in 2018 (EIA 2017a). This marks a sharp turn from earlier EIA projections putting the US on a firm pathway towards rising imports. The EIA's 2008 Annual Energy Outlook had still warned that by 2030, America would need to see 4.5 trillion cubic feet or 127 bcm of additional LNG intake in order to compensate for the declining domestic output and satisfy growing demand at home (EIA 2008, 94). The USA therefore started building up regasification capacity for LNG import, which raised concerns in Washington's security circles. Questions arose around increasing dependence on external producers, as well as rising costs for consumers (Urban 2008). Debates also centred on the possibility of the members of the Gas Exporting Countries Forum (GECF) – notably Russia, Qatar and Iran – forming a potential 'gas cartel' which could strangle the

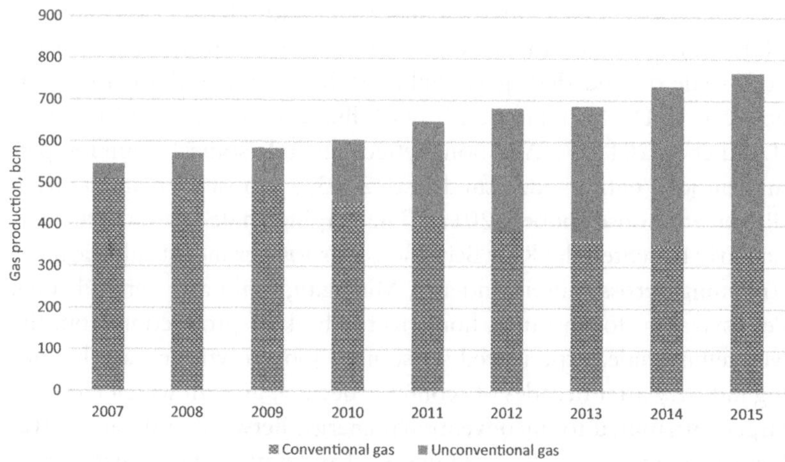

Figure 2.1 US unconventional gas production, bcm per year
Source: (EIA 2016b, 2017f); author's own calculations

country as the Organization of Petroleum Exporting Countries (OPEC) did with oil in the 1970s (International Herald Tribune 2008).

Shale gas turned the tide. With total gas output standing at 760 bcm a year in 2015, the USA emerged as the world's leading gas producer, surpassing Russia (570 bcm) and being poised to retain the top spot for years to come (BP 2016). With estimated reserves of 1400 Tcf or 39 648 bcm of natural gas, estimated as recoverable at 2016 Henry Hub spot prices, the country will be able to keep production levels on an upwards trajectory for decades to come (IHS Energy 2016). US shale gas production also proved remarkably resilient to challenges arising from a low oil price environment impacting on shale oil output (which adds associated gas to the balance), thanks to a continued learning curve, cost reduction and a favourable bankruptcy law helping to clear liabilities What had started off as a niche for Texan wildcatters turned into America's next energy frontier – and a multi-billion dollar industry – within a mere decade.

The economic effects of the unconventional gas industry are significant. Benefits come in the shape of consumer surplus for residential, commercial, industrial and electric power customers; as producer surplus thanks to a strong increase in natural gas output; and through local economic effects such as royalty payments and additional state income from severance taxes, fees and levies (Hausman and Kellogg

2015; Mason, Muehlenbachs and Olmstead 2015). According to McKinsey estimates, the unconventional energy industry (comprising both shale oil and shale gas) could add between 2 and 4 per cent, or between USD 380 billion and 690 billion, to the US GDP by 2020 (Lund et al. 2013, 7). Additional effects include some 1.7 million permanent jobs nationwide (Lund et al. 2013, 7), most in manufacturing (PricewaterhouseCoopers 2011). The surging shale gas industry particularly benefited the Rust Belt, the USA's former industrial heartland stretching across Illinois, Indiana, Michigan, Ohio and Pennsylvania. Pennsylvania, for instance, home to the bulk of production from the Marcellus Shale, experienced substantial job growth and a rebounding industry after decades of economic decline: growth which has been largely attributed to unconventional energy. Between 2007 and 2012, Pennsylvania saw employment in the oil and natural gas industry grow by 259.3 per cent, which was flanked by wage increases of 36.3 per cent in the sector (US Bureau of Labor Statistics 2014). Similar trends can be observed in Ohio and Michigan, as both states saw a rise in shale gas production. While some recent studies added a note of caution to overly optimistic estimates on the positive effects for the deindustrialized Rust Belt (Hoy, Kelsey and Shields 2017), the surge in shale gas triggered an economic 'transformation spreading across the heartland of the nation' (New York Times 2014a).

As a corollary of the surge in American gas output, domestic natural gas prices faltered. Since 2009, Henry Hub, the US marker, has hovered around 3 US dollars per million British thermal units (USD/MMBtu – see Figure 2.2), which benefits consumers and industry, lowering their energy bills. In fact, US consumers have come to enjoy some of the lowest gas price levels in the world. Asian consumers saw LNG import prices soaring close to the 20 USD/MMBtu mark in the aftermath of the Fukushima nuclear disaster, and European pipeline gas were adjusted upwards as a function of rising oil prices, reaching 10 USD/MMBtu in 2011. The Henry Hub spot price, by contrast, remained at a fraction of that, and occasionally even dipped towards the 2 dollar mark. Even though Asian prices saw another downwards trend towards the 10 USD mark as of 2014, the well-supplied US market remained significantly more price-competitive and ensured consistently low energy costs for households and businesses.

What's more, because of its increasing competitiveness, natural gas started replacing coal in the power sector. Between 2000 and 2012,

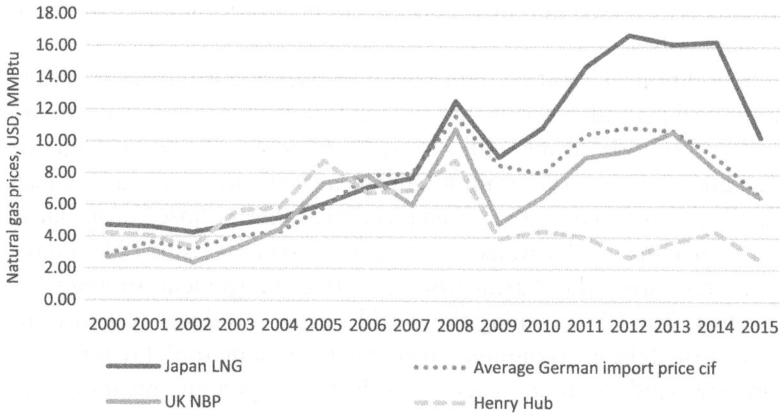

Figure 2.2 Natural gas prices 2004–2013, USD/MMBtu
Source: BP (2016)

gas-fired electricity production capacity nearly doubled in the USA (EIA 2011a). This trend, which accelerated further as of 2010, is primarily a function of gas outpricing lignite and hard coal (in addition to policies such as President Obama's Clean Power Plan and a surge in renewables depressing coal demand). Eventually, 2016 marked the year when natural gas took the top spot in US power production, surpassing coal for the first time in US history. Natural gas now accounted for 33.8 per cent of the country's power supply, and coal for 30.4 per cent – a shift in fuel shares which, according to the EIA, clearly reflects longer-term trends (EIA 2017c). Because natural gas comes with a lower carbon footprint compared to coal, the strengthened position of natural gas sharply brought down CO2 emissions in the country.[1] By 2016, the cumulative CO2 emissions of the USA had fallen to 1992 levels (EIA 2016a, 2017e). In the run-up to the Paris Agreement of 2015,

[1] It is important to note that although natural gas is a 'cleaner' fossil fuel compared to coal, shale gas comes with higher greenhouse gas (GHG) emissions than conventional natural gas. Hultman, Rebois, Scholten and Ramig (2011) find that on a lifecycle basis, the GHG emissions of shale gas are 5 per cent higher than those of conventional gas, but still remain 42 per cent lower than those of coal. By contrast, Howarth, Santoro and Ingraffea (2011) and Howarth (2015) argue that on a lifecycle basis, shale gas comes with even higher GHG emissions than coal. The latter studies have been contested and the exact GHG 'footprint' of shale gas remains subject to an ongoing scientific discussion.

this allowed the US administration to claim a leadership role in global climate policy.

In addition to boosting domestic industry, revitalizing economically ailing regions and lowering the US CO2 footprint, America's newfound gas (and oil) also came with a security premium. To be sure, import dependence for the USA traditionally meant reliance on Canada, which clearly did not come with challenges comparable to those Europe faced with Russian gas. But America's trajectory towards rising LNG intake levels had given the Washington security establishment growing discomfort. The US becoming the world's prime producer of hydrocarbons turned the geoeconomics of energy to Washington's favour again. On one hand, domestic energy production represents an important asset in the global struggle for economic supremacy. Not only was the USA among the first countries to rise again from the Great Recession that started in 2009, but the shale-induced boost in industrial competitiveness also directly benefited it internationally, as it allowed the fostering of strategic trade agreements largely on American terms, both in the Asia-Pacific and across the Atlantic. Europeans toying with the idea of an 'energy chapter' in the – now moribund – Transatlantic Trade and Investment Partnership (TTIP) also points to the USA gaining influence in international energy trade patterns (The Guardian 2016a). This extends to a grand-strategy reading of potential US gas exports to EU allies, with a view to curbing Russian influence on European markets and, by extension, domestic politics. With this, the shale boom translates into a means of US statecraft and foreign policy (O'Sullivan 2013).

In all, the US 'shale revolution' came with a triple premium: low energy prices and an economic boost; a reduced carbon footprint underpinned by simultaneous economic growth; and a gain in international security and geoeconomic standing. Unsurprisingly, many countries around the globe sought to emulate and replicate the US success story. This applies particularly to CEE, a region we will turn to next.

2.2 Energy Security in Central Eastern Europe: Russia and Gas Import Dependence

Ever since the Soviet Union inked its first gas trade deal with West Germany and Italy in the late 1960s and early 1970s, European energy security has been on the policy agenda – although arguably more so in

Washington, DC than in Western European capitals. Early on, the US leadership started to worry about Europe's supply security (New York Times 1981), a theme that remained prominent all the way through the 2006 and 2009 Ukrainian–Russian gas crises, which we shall discuss later. Western European plans to source Soviet gas also led to growing concerns in Washington about US allies becoming politically vulnerable in the face of potential gas cut-offs (Hardt and Gold 1982).

Western Europe striking gas deals with the Soviets came with its own logics, of course. West Germany's energy contracts with Moscow have to be seen in the context of *Ostpolitik*, chancellor Willy Brandt's politics of détente with the Soviet Union (Stent 1981). Brandt's 'rapprochement through trade' approach sought to deepen mutual economic interdependency between the West and the Soviet bloc, including through gas trade. The East–West gas deals of the 1970s also came against the backdrop of strong economic growth in Europe for most of the postwar period, which meant increasing energy needs in terms of both oil and gas. The Soviets were prepared to satisfy those needs in exchange for hard Western currency. An emerging East–West gas trade also gave a boost to the European steel and manufacturing industries, as their technology leadership helped them secure lucrative pipeline contracts. Germany, Italy, Austria and France therefore went ahead and pushed for closer energy ties with Moscow, regardless of Washington warning against such a move. This resulted in the development of some of Siberia's major gas reserves, including the giant Urengoy field, and the construction of a long-distance pipeline system bringing West Siberian gas to the Soviet Union's western border – including the now disputed Urengoy–Pomary–Uzhgorod pipeline – for further distribution within non-Communist Europe. By 1990, Western Europe sourced some 49 bcm of gas, or 22 per cent of its consumption (Smeenk 2010, 122), from what Ronald Reagan famously called the 'evil empire'.[2]

When the Iron Curtain fell, Central and Eastern European countries added their distinct dependency pattern to the European gas conundrum. As members of the Council for Mutual Economic Assistance (COMECON), these countries had been the recipient of gas (and oil) supplies from the Soviet Union in exchange for industrial or manufactured goods: the COMECON model of economic

[2] For a comprehensive historical treatise of 'red gas' and the emergence of East–West energy relations, see Högselius (2012).

Table 2.1 *Russian import dependence of select EU countries, per cent and bcm, 2011*

	Consumption	Imports from Russia (net)	Share
Bulgaria	2.2	2.2	100%
Estonia	0.6	0.6	100%
Slovakia	5.6	5.5	98%
Poland	17.2	10.0	59%
Hungary	11.5	5.2	45%
Romania	12	3.0	25%
Germany	77.6	31.1	40%

Sources: BP (2016), IEA (2012b) and KPMG (2012)

specialization. While Poland and Romania had non-negligible domestic gas and oil production, Eastern Germany, Czechoslovakia and Bulgaria did not, which essentially meant these countries retained a complete dependency on Russian gas imports post-1990. What's more, the pre-existing energy infrastructure physically hardwired an East–West pattern into Eastern Europe's import network. Pipelines were designed and built to bring gas from Western Siberia into Communist Eastern European countries, feeding into national distribution networks there. With twelve Eastern European countries joining the EU in three consecutive rounds of enlargement in 2004, 2007 and 2013, it is therefore not only the overall size of the European gas market that grew (to stand in 2015 at more than 400 bcm; BP 2016), but also Russia's dominant position in the EU's import portfolio (representing 39.7 per cent of EU consumption in 2015; Eurostat 2017). (For 2011, the year roughly marking beginning of the period of this investigation, see Table 2.1.)

Besides significant infrastructure investment that would be needed to overcome this import dependency pattern, the COMECON legacy also gave rise to important incumbent stakeholders in Eastern Europe: national utilities, now in charge of managing imports and distribution. These were mostly state-owned, and often still are, as illustrated by Poland's PGNiG, Bulgaria's Bulgargaz and, in the case of the Former Soviet Union (FSU), Ukraine's Naftogaz. National energy champions not only dominate domestic markets, as they typically operate gas imports from Russia, which come in the shape of bilateral LTCs (see later), but in some instances also control the downstream segment. For governments, they constitute an important means of ensuring national energy supply and economic security. Put differently, it is the public utility model that primarily informs CEE energy governance still, rather than the liberal market model – and it is this model which, as we shall argue later, hinders change. As a corollary, energy regulation tends to be geared towards serving the needs of a few dominant state actors, rather than private corporations. Regulatory legacy hardwires the incumbents' dominant role into the domestic energy system, a pattern that has proven cumbersome to overcome in Eastern Europe, even against the backdrop of a determined pro-market push during and after EU accession. Top-down price regulation and heavy-handed state interference remain part of the governmental toolbox in energy policy. Mergers and acquisitions in the energy sector also retain a strongly strategic character in the region (Butler 2011), and some countries have even seen a reversal of earlier liberalization and privatization policies. A case in point is Hungary, which, as of 2010 effectively renationalized gas storage and increased the state ownership share of MOL, the national oil and gas company (Euractiv 2012a).

What's more, the Eastern European gas conundrum comes with a distinct infrastructure challenge. The dissolution of the Soviet Union and the (re-)emergence of Belarus and Ukraine as sovereign FSU states impacted on the export network of Gazprom, Russia's state-owned company succeeding the Soviet Ministry of Gas Industry. What had been designed as an integrated export pipeline system during Soviet times now became at least partially owned and operated by various USSR successor countries and their (state) companies. A case in point is the Urengoy–Pomary–Uzhgorod pipeline, home to roughly 90 per cent of Russia's gas exports to Europe in the early 1990s, which ended up crossing Ukraine before feeding into today's Slovakia. In fact, natural gas transit became a contentious issue and led to repeated tensions between Russia and Ukraine, as well as between Russia and Belarus. These tensions are epitomized by, and in fact culminated in, the Russian–Ukrainian gas disputes of the winters of 2006 and 2009. In both instances, Gazprom stopped its gas supplies to Ukraine. Yet, while the 2006 dispute lasted only a few days, the 2009 crisis ended up stretching over almost two weeks. In the latter incident, Russia also not only halted volumes destined for Ukrainian consumption, but cut gas exports through Ukraine altogether (Stern 2006; Stern, Pirani and Yafimava 2009). This affected a total of eighteen EU countries, plus Moldova, further down the pipeline. Given their higher import dependency rates and a lack of alternative sources compared to Western Europe, it was exclusively countries in Eastern and South Eastern Europe that were affected, and whose socio-economic life was in some instances brought to a halt (Kovacevic 2009).

At their core, both crises were about disagreements over gas prices and volumes, transit fees and outstanding debt. A less technical view, however, suggests complicated political economy dynamics arising from export pipelines transiting a third country's jurisdiction. Ukraine's sitting on a 'geographical monopoly' (Stevens 2009) over Russian gas exports (and European gas imports) enabled Kyiv to retain relatively low gas prices after the fall of the Iron Curtain and to siphon off gas destined for European customers, as it did during the 2006 crisis. For Ukraine, Gazprom's efforts to end the prevalent contractual arrangement and to establish a different pricing regime amounted to letting go of what were effectively monopoly rents, and culminated in the gas disputes. What makes these disputes very political, however, is their timing. In 2005, Ukraine had just seen the Orange Revolution, resulting in the ousting of President Viktor Yanukovych, a Moscow ally, and bringing about pro-Western political change. The new president, Viktor Yushchenko, had run on a reformist platform, sought Ukrainian NATO membership and refused to guarantee the Sevastopol lease of Russia's Black Sea fleet. This made him a potential threat to Russia's ambitions to retain Ukraine in its geopolitical orbit. Observers have therefore linked both gas crises to broader Russian foreign policy goals (Baev 2008; Nygren 2008).

The Russian–Ukrainian gas disputes impacted significantly on Gazprom's reputation as a reliable supplier of gas to Europe. At the same time, however, they highlighted the fact that Russian energy policy cannot be entirely separated from Moscow's foreign policy goals. Between 1991 – the year marking the end of the Soviet Union – and 2005, Larsson (2006) counts fifty-five incidents of Russian energy cutoffs or coercive pricing, of which more than forty are attributed political motivations. More than forty supply cuts are reported to having targeted the Baltics and Commonwealth of Independent States (CIS) countries (Larsson 2006, 262). In other words, the 2006 and 2009 crises can also be interpreted as two incidents in a longer series of events. This ties into a broader debate on energy – and particularly gas – as a means or end of the Kremlin's statecraft, and on Russia as an 'energy superpower' (Cohen 2009; Goldthau 2008; Rutland 2008). Stulberg, for instance, elucidates how Moscow uses political levers to secure control over Caspian energy resources and Eurasian energy trade more generally (Stulberg 2005, 2008). Abdelal stresses the role of companies in this context, linking energy corporations and realpolitik (Abdelal 2013). Conversely, Russia securing an eventual extension of the Sevastopol lease in 2010 came on the back of price cuts for Ukrainian gas contracts (The Guardian 2010), a longstanding practice in which energy pricing forms part of the Kremlin's foreign policy toolbox in regards to former Soviet states (Astrov 2011; Balmaceda 2008).

In addition to damaging Russia's reputation as an energy player, the gas disputes tied into broader national security concerns, particularly in Eastern Europe. Against the backdrop of their half-century-long experience as the Soviet Union's satellite states, many Eastern European capitals followed Moscow's energy and foreign policy ambitions with deep-seated distrust. Fears centred on Russia seeking to reinstate its rule over Eastern Europe and Moscow meddling in domestic affairs. In the eyes of regional leaders – but also many observers in Western capitals – Moscow's alleged use of energy supplies for political ends during the gas disputes had to be seen in this context. To them, mounting evidence suggested an increasingly assertive behaviour in Russia's 'Near Abroad'. This included Russia supporting or fostering breakaway movements in Transnistria, South-Ossetia, Abkhazia and Nagorno-Karabakh – and, as of 2014, the Donbas and Crimea; instigating domestic turmoil, for instance in Estonia over the removal of a Soviet war memorial in 2007; or even engaging in outright conflict, such as in the 2008 Russia–Georgia war. Against that backdrop, Moscow's foreign policy had come to be perceived as aggressive and revisionist by the end of the new millennium's first decade (Krastev 2014).

Because of their strategic nature, energy affairs are therefore typically cast in strongly geopolitical terms across Eastern Europe, rather than in economic ones. Gas contracts and infrastructure projects in particular tend to be read against the backdrop of the region's history, especially the role Soviet Russia played there in the twentieth century. Historical sensitivity also clearly exists regarding Germany, as the memories of the Nazi military machine waging its campaign of destruction are quite present in in the region still. This point is epitomized by the Nord Stream pipeline project. The 55 bcm gas link between Russia's Vyborg and Germany's Greifswald became strongly disputed for allegedly fostering ties between Moscow and Berlin and renewing a Russian–German political relationship that had proven detrimental for Eastern Europe in the past. The project was likened to the infamous Molotov–Ribbentrop Pact between Stalin's Soviet Union and Nazi Germany, dividing up Eastern Europe and preceding the German invasion of Poland. Nord Stream 2, the planned expansion, triggered similar reactions, with nine Eastern European heads of states even signing a joint declaration against the project, for its alleged negative geopolitical consequences for the region. In turn, Gazprom's gas transit through the Yamal pipeline, crossing Poland, is regarded as an insurance policy against adverse Russian (and German) behaviour. For similar reasons, Eastern European leaders have been vocal in calling to retain Ukrainian gas transit to Europe, countering Russian efforts to circumvent Ukraine once a current contract expires in 2019. Tellingly also, Lithuania's first LNG terminal, bringing Norwegian gas to the country, was named 'Independence', and cheered for its energy security benefits for the Baltics more broadly (New York Times 2014b).

While it is debatable whether individual pipeline projects indeed constitute a geopolitical threat to the region (Goldthau 2016a), it would be short-sighted to dismiss Eastern European energy security concerns as completely unfounded. As the October 2014 'stress tests' – initiated by the European Commission at the height of Russian–Ukrainian tensions over Crimea, Donbas and Luhansk – suggested, many countries in the region would indeed suffer from a supply shock in gas. Even in the event of member states sharing the burden and acting in the spirit of solidarity, the Commission found in its report, the impact of a six-month disruption of Russian gas deliveries on the Baltic countries, Hungary or Bulgaria would be 'non-negligible'. Short of mutual solidarity, these countries would even end up seeing 'serious supply shortfalls' (European Commission 2014d, 6). The stress tests also highlighted the vulnerability of the 'energy islands' in the Baltics and South Eastern Europe. Overall, the Commission report revealed that although progress had been made compared to 2009, the region remained vulnerable to supply shortages and their consequences. Unsurprisingly, therefore, Russia's 2014 annexation of

Crimea reconfirmed longstanding Eastern European perceptions of Russia as a potential security threat, and added to concerns around high gas import dependence.[3]

A final element in the Eastern European gas conundrum pertains to import prices. Wholesale prices across Eastern Europe have been reported to exceed prices charged in Western Europe. A snapshot for the third quarter of 2012, for instance, shows an average import price of 24.9 EUR/MWh for Russian gas sold at the German border, whereas the price for Poland was 29.6 EUR/MWh, for Hungary 32 EUR/MWh, for Lithuania 39.6 EUR/MWh, for the Czech Republic 40.2 EUR/MWh and for Bulgaria 40.5 EUR/MWh (European Commission 2012b, 18). Note that this is after the EU introduced its 2009 'Third Energy Package' (see later), mandating a deeper integration of European gas markets, which started to align prices more closely across the bloc. While price levels change and adjust over time as a function of, among other things, oil price movements (and, more recently, spot market developments), the general pattern of Eastern Europe paying higher import prices holds over the years all through 2015. This divergence between Eastern and Western European wholesale gas prices prompted Eastern European political leaders to deplore 'unfair' and 'political' pricing on part of Gazprom – claims that, in 2012, led the EU Commission to launch an anti-trust procedure against the Russian monopolist (European Commission 2012a). It also became an economic policy issue, given the significantly lower national income in the region. With a per capita GDP remaining below EU average – 69 per cent in the case of Poland, 57 per cent in Romania and 47 per cent in Bulgaria – comparably high gas import prices and their welfare effects became a concern on their own, independently of import-related security aspects.

Overall, energy security continues to feature prominently on the Eastern Europe policy agenda. To be sure, Western European countries such as Germany and Italy also have high import dependency ratios, with Russia representing an important source of gas supplies. But it is a more pronounced exposure to supply risks, historical sensitivities regarding Moscow, economic security concerns and legacies in infrastructure and regulatory governance that countries joining the

[3] Note that scholars warned against establishing a direct link between Crimea and energy geopolitics (Boersma and Goldthau 2014).

Table 2.2 ***Shale gas reserve estimates and annual consumption, bcm, select countries***

	Technically recoverable resources (2013)	Demand (2013)	Years of nominal consumption
Europe	13 310	432*	30
Poland	4191	16.6	252
France	3880	43.1	90
Romania	1444	11.3	127
Bulgaria	481	2.6	185

Sources: EIA/ARI (2013), BP (2016), own calculations. * EU-27

EU in 2004 and 2007 bring to European energy policy. Although these countries by no means represent a 'united front' in EU energy affairs (Nosko and Mišík 2017), their energy policy agendas clearly come with a strong geopolitical spin. In a nutshell, (energy) economics are trumped by high (energy) politics.

Unsurprisingly, therefore, the shale revolution unfolding in the USA was met with great interest (and a good dose of jealousy) in Eastern Europe (New York Times 2010). Shale gas was known to be existent across the region, but with detailed estimates being unavailable, and the technology lacking, only a few countries, such as Poland, were actively pursuing the new energy frontier. It was in the wake of a 2011 EIA-commissioned study on unconventional hydrocarbon resources that this interest turned into outright excitement in Warsaw, Sofia, Riga and Bucharest. The EIA study, the first to comprehensively survey global reserves, suggested that unconventional gas could be found in many parts of Europe. It nominally put Europe's 'technically recoverable' shale gas reserves at 624 tcf or 17671.68 bcm – more than forty years of total European consumption. Eastern European resources (including Ukraine and Kaliningrad) were estimated at 252 tcf or 7136.64 bcm and represented a large share of overall reserves (EIA 2011b). Poland and the Baltics were named as potentially promising basins for shale gas. A 2013 follow-up assessment detailed additional reserve estimates in Eastern Europe, including in Romania and Bulgaria; confirmed Polish shale reserves as the largest in Europe, ahead of France's; and put total technically recoverable resources in shale gas at 470 tcf or 13310.4 bcm (Table 2.2) (EIA/ARI 2013).

To be sure, even the more rigorous 2013 EIA numbers were mere estimates, and in some instances they saw significant adjustments, mostly downwards (see Chapter 4). Yet, they inspired Eastern European countries to seriously consider developing domestically available resources with a view to yielding a 'triple premium' similar to that in the USA. Replicating the American success story, it was hoped, would enable some troubled CEE countries to put an end to decade-old security-of-supply woes and would thus be a 'game changer' (Politico 2013) in regional (energy) geopolitics. The notion of 'energy independence' from Russia captured the imagination of political leaders and ordinary people alike. What's more, it was hoped that domestically available gas would bring down prices and give national economies a competitive edge. In the context of the post-2008 Great Recession, this would also boost much-needed jobs and taxable income in the region. A 2013 study carried out by Poyry Management Consulting and Cambridge Econometrics suggested that in a 'shale boom' scenario, the EU could benefit from a EUR 3.8 trillion GDP injection between 2020 and 2050, and an additional 1.1 million jobs (Pöyry 2013). Knock-on effects, the study suggested, would particularly benefit energy-intensive industry through lower gas prices. While these estimates clearly are debatable, they supported the broader sentiment that shale gas should be given a shot. In short, in the view of regional leaders, Eastern Europe was ready for the 'dash for shale'.

2.3 Shale in the Context of European Gas Market Dynamics

At this point, it is important to note that energy security concerns are not unique to gas. In fact, the most prominent case of energy and politics coming together was probably the 1973 oil shock. Punishing Western countries for their support of Israel in the Yom Kippur war, Arab oil producers stopped crude shipments to the USA, Canada, Japan, the Netherlands and the UK, and unilaterally quadrupled the oil price. The 1973 events famously marked the end of a decade-long period of post-war economic growth and brought oil back to international politics. Yet, energy security concerns in oil are hardly comparable to those in gas. Because the 1973 oil shock affected most Western countries in one way or another, the OECD responded by way of creating a joint institution for emergency management, the International Energy Agency (IEA), which was tasked with buffering supply shocks and coordinating its member countries' policies on strategic petroleum stocks (Kohl 2010). Moreover, with oil becoming a truly global commodity in the 1980s, the threat also no longer consisted in potential physical disruptions of petroleum supply – not even from the Soviet Union, which ended up exporting much of its crude to Europe in return for hard currency.[4] Instead, price hikes and volatility became the prime challenges in oil market governance (Goldthau 2011a). Today, although undoubtedly not representing 'just another commodity', oil's fungible nature, coupled with a highly complex and globally integrated market, makes crude prices essentially a function of supply and demand. With this, energy security in oil by and large rests on mitigating or preventing market failure; that is, addressing questions of information and market transparency, producer collusion and spillovers from non-market events (Goldthau and Witte 2009).

Natural gas, by contrast, by and large remained a regional play, a function of the commodity's fugitive nature and the resulting infrastructure requirements: pipelines. As a corollary, distinct Eurasian, North American and Asia-Pacific markets persist, each characterized by its own pricing and trading patterns. Interregional gas trade is projected to pick up considerably in the years to come, thanks to LNG. Yet, although LNG has seen significant growth rates since the early 2000s, the bulk of the gas trade will remain pipeline-bound until 2040 (IEA 2016). Even then, Europe might well trail behind broader trends in global gas trade, given an import infrastructure that is strongly reliant on long-distance upstream pipelines.

Moreover, while national-level measures and institutional arrangements for dealing with supply emergencies are in place, notably in the shape of gas storage, they typically remain rather underdeveloped on a regional level. Besides nation states eagerly watching over their national prerogatives in energy policy, there also exists a collective action problem when it comes to putting in place costly (national) infrastructure serving a regional public good: energy security (Goldthau 2011b). A case in point is the EU's 2010 gas security directive (European Parliament and the Council 2010c), which took two major gas crises to come into effect. Energy security in gas, therefore,

[4] For a discussion of the Soviet oil economy, see Gustafson (1998).

still remains a matter of physical supply, with inevitable price effects consequent on any shortage in supply.

In light of this, shale gas could arguably impact on the European energy situation beyond CEE. First, it could be a welcome addition to the European gas balance. Given comparably low indigenous production, European gas markets overall remain highly dependent on imports from external suppliers. In 2015, EU gas output stood at 120 bcm, which compares to around 400 bcm of domestic demand (BP 2016). What's more, EU production has been on a steady decline since 2000. This is to do with geology (i.e. maturing fields, for instance in the UK's North Sea), but is also a result of political decisions, such as the Netherlands introducing a cap on production from the important Groningen field, as a result of which output decline has accelerated sharply (Reuters 2016a). Norway, a key European supplier tied to the EU as part of the European Economic Area (EEA), will also see a peak in production around the 2020s, with output falling to 84 bcm in 2040 (IEA 2015). This trend is projected to continue all the way through the mid-century. Per IEA estimates, indigenous EU production will be down at 92 bcm a year by 2040, which amounts to an almost 50 per cent decline compared to 2013. The European Network of Transmission System Operators for Gas (ENTSO-G) suggests a contraction of up to 68 per cent through 2035 (ENTSO-G 2015). The EU Commission's own projection are only slightly more optimistic, putting domestic output levels at 95 373 ktoe or 103 bcm in 2040 (European Commission 2014i).

EU demand, by contrast, has been flattening out for years, due to increasing energy efficiency, the fallout of the economic crisis starting in 2008 and competition from renewables and cheap coal imports, mainly from the USA. Going forward, this demand trend is not likely to be reversed. In its updated impact assessment for the Energy Roadmap 2050, the EU Commission expects an annual gas demand of 397 669 ktoe or 429 bcm in 2040 (European Commission 2014i). The IEA puts less weight on renewables in the EU energy mix, and in its New Policies Scenarios lands at 466 bcm by that time (IEA 2015). Eurogas, the industry association, assumes a corridor of demand of 437–585 bcm by 2035, with actual consumption being a function of whether public policies are more or less favourable for natural gas as compared to other fuels (Eurogas 2013). Except for Eurogas estimates, EU

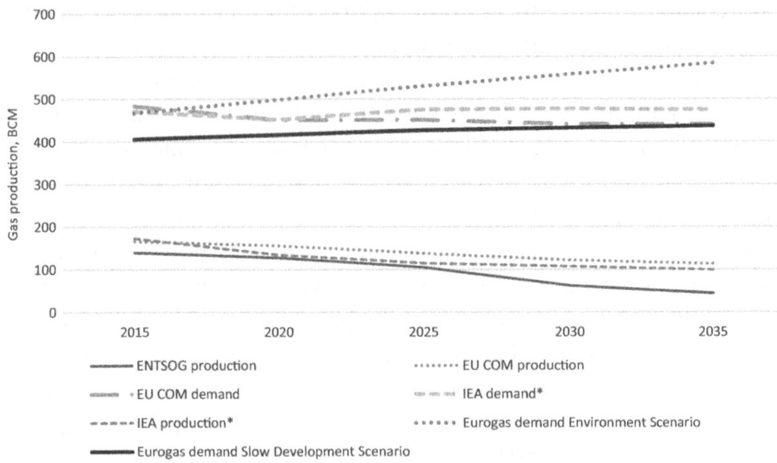

Figure 2.3 EU gas balance
Sources: ENTSO-G (2015), European Commission (2014i), IEA (2015) and Eurogas (2013). * 2013 data for 2015

consumption in 2035 and 2040 is therefore likely to stand roughly where it was in 2015.

Still, these trends suggest a growing gap between projected demand and domestic production (Figure 2.3). With this, the EU's import dependency ratio is set to grow as well. In the IEA assessment, imports will meet 83 per cent of EU demand by 2040. BP estimates the share of imports at almost 80 per cent of EU demand by 2035 (BP 2017). Shale gas, it has been suggested, could therefore at least stabilize current import rates by making up for some of the projected decline (Inmam 2016), or even decrease the overall import dependency ratio – by some estimates, to as low as 62 per cent (Pöyry 2013).

A second aspect pertains to the degree to which domestic shale gas production could support EU-level efforts to build a common EU energy market. The traditional gas market model, in both Eastern and Western Europe, was built on LTCs tying a producer (such as Norway's Statoil or Russia's Gazprom) and a consumer (such as Germany's Ruhrgas or Italy's ENI) in to a long-term relationship.[5] LTCs would run fifteen or more year, and defined volumes of gas supply

[5] The UK, a longstanding gas producer, is an obvious outlier here, having liberalized its gas market in the 1980s.

that the producer had to deliver over that time and which, conversely, the consumer had to take off – or pay (hence, the term 'take-or-pay agreement'). LTCs would also typically tie the price of gas to that of oil.[6] Moreover, despite decade-long European economic integration, gas markets had effectively remained national until 1992, with little or no cross-border trade to neighbouring countries happening within the bloc. This situation was to change with three consecutive 'Energy Packages' put forward by the European Commission, which were intended to push for a more market-based approach by opening up national markets and integrating them across the EU. The Commission's approach was a gradual one, starting with the rather light-touch 1998 Directive (European Parliament and the Council 1998) and ratcheting up market liberalization with two follow-up directives in 2003 (European Parliament and the Council 2003) and 2009 (European Parliament and the Council 2009d). Focus was placed on fostering market competition by way of demanding legal and ownership unbundling of gas transport from gas trading services, establishing independent regulators and ensuring non-discriminatory third-party access. The Commission also put an end to destination clauses limiting gas resale in LTCs, notably with Russia's Gazprom (European Commission 2003).

In this way, as observers have noted, EU energy policy by and large followed the regulatory state model (Lodge 2008; Majone 1994), focusing on 'making markets and making them work' (Goldthau and Sitter 2014, 2015a). The tools employed were regulatory ones; that is, directives, regulations and competition policy; and the basis for the policy was the Commission's mandate to foster economic integration. This is not to say that the EU was not aware of the broader geopolitical environment pertaining to gas markets. Regulatory instruments as applied were often at the crossing between avoiding the market dominance of Gazprom and catering to political goals such as curbing Russian influence (Goldthau and Sitter 2015b). But, because of the EU's limited policy toolbox, its origins as an economic integration project and, indeed, the institutional set-up of the Commission as the main actor behind the pro-market push, EU-level efforts to address energy security concerns essentially followed a liberal blueprint. Besides improving what EU Commission Vice President Maroš Šefčovič refers to as the

[6] For details on gas pricing history, see Energy Charter Secretariat (2007) and Konoplyanik (2010).

market 'software' – energy regulation – emphasis was also placed on upgrading the energy 'hardware' in the shape of strategically important interconnectors, pipelines bringing additional Caspian gas to the EU market and LNG terminals sourcing from an emerging global market (European Commission 2016d).[7] As part of the EU's efforts to manage market risks, a security of supply regulation was introduced (European Parliament and the Council 2010c) – essentially, a regulatory state answer to market failure.[8]

To be sure, change did not come quickly, and the EU's efforts were not always crowned with success. Nabucco, the EU's favourite pipeline project, aimed at linking the bloc to Caspian gas reserves, had to give way to the much smaller TAP project and left South Eastern European countries in a persistent supply security limbo; strategically important intraregional gas infrastructure projects such as the 'Vertical Corridor' – consisting of the Interconnector Greece–Bulgaria (IGB) and the Romania–Bulgaria Interconnector (IBR) – saw significant delays due to regional governments dragging their feet; and several infringement procedures against member states for failure to implement EU energy regulation not only delayed market integration but also a suggest strong material interest in the status quo among incumbent actors. And yet, the EU's push towards a liberal energy regime altered EU gas markets. By 2016, interconnectors had come to link most national markets, spot trading and gas-on-gas competition had taken over as the most common form of pricing and markets had become significantly more competitive.

The EU's efforts to remodel a decade-old incumbent system were, however, also greatly helped by the shale revolution unfolding on the other side of the Atlantic. US shale gas production started to surge just around the time Qatar became the world's largest LNG exporter. Qatari LNG, which had been earmarked for steeply growing US demand, found the American market increasingly saturated with domestic production and had to find sale elsewhere. At the same time, the 2008 financial crisis put the world economy into a deep recession, depressing gas demand not only in the USA but also in Europe, now the destination of choice for Qatari LNG. The EU saw consecutive demand cuts of 7 per cent in 2010 and 10 per cent in 2011 (BP 2016).

[7] For a detailed discussion of EU energy policy initiatives, see Goldthau (2013).
[8] See also Boersma (2015).

The combination of faltering global gas demand and the US market effectively closing down for gas imports led to a 'perfect storm' for incumbent suppliers to the European market, such as Russia and Norway (Hulbert and Goldthau 2013). With global LNG cargoes being redirected to Europe, spot markets there became soft. Facing a growing gap between faltering spot prices and oil-pegged LTCs, European utilities such as Germany's E.ON Ruhrgas, Italy's Eni and France's GDF Suez found themselves out of the money, as their business model was tied to LTCs. As a result, they turned to their suppliers and started to push for contractual adjustments, demanded rebates and revisions of their LTCs' incumbent pricing structures and even looked to arbitration. Norway's Statoil jumped first and granted rebates, and reluctantly Gazprom followed suit. Both Statoil and Gazprom also agreed to revisit contractual pricing formulas and introduced elements of spot pricing in their running LTCs. As of 2012, the majority of the gas volume sold in Europe was indexed to the spot market rather than oil, and by 2015 gas-on-gas competition made up 64 per cent (IGU 2016).

As a corollary, gas markets that had so far remained very regional in nature started to sync. LNG trade emerged as the vehicle to link the Asian and European markets, leading to price alignments on spot markets. Still, however, important regional characteristics persist. Natural gas is far from being the global and liquid market represented by oil. Spot markets remain comparably small, prices are therefore volatile and LNG cargoes are partially tied into LTC arrangements, reflecting the sector's large upfront costs. The transition towards a fully integrated international gas market will take time, and particularly in the context of Eurasian markets, this will require management (Goldthau 2012a).

This brings us back to the prospects of European shale gas supporting EU energy market integration. While much progress has been made, the policy approach of leveling the playing field for external producers faces limits, given a small number of players supplying the EU market. With effectively three non-EU producers – Russia (40 per cent), Norway (37 per cent) and Algeria (7 per cent) – covering the bulk of gas imports in 2015, and LNG making up an additional 13 per cent (European Commission 2016f), competition and market liquidity may still end up facing limitations. Hindering competition from within, national governments fail to implement pertinent EU regulation, falling short on energy infrastructure investment goals and seeking to retain price regulation as a means to secure domestic political support. As a result, differences in pricing structures remain significant across European markets, with the north-eastern market being almost fully hub-based and just above half of CEE market volumes being operated on gas-on-gas competition. The Mediterranean, by contrast, remains firmly tied to oil price indexation, representing two thirds of the market, and in South Eastern Europe price indexation and regulated pricing structures prevail (IGU 2016, 41ff.). Domestically available molecules – in the shape of both conventional and unconventional gas – would strongly underpin the 'market-making' efforts on the part of the European Commission, increase market liquidity and competitive pressure, and thus help change the market from within.

With this, we briefly turn to the question of how shale gas ties into the broader EU apparatus pertaining to energy and environmental regulation.

2.4 EU Energy Policy-Making: Between Multilevel Governance and National Prerogatives

It is often said that the EU is a messy polity. And, indeed, the specific character of the EU, its complex institutional set-up and the way it sets policies have given rise to an entire scholarly industry. It is far beyond the scope of this book to delve into the working of the EU or detail its institutional intricacies, for which pertinent volumes in EU studies are much better positioned.[9] For the purpose of this discussion, probably the most sensible way to approach shale gas is to ask what the EU can do in energy and the environment. In this context, the notion of subsidiarity is an important one. Subsidiarity, the principle for governing the EU, essentially restricts supranational – that is, EU-level – action to areas where Brussels enjoys exclusive policy competence. The subsidiarity principle is based on the idea that regulation and policy should sit and be implemented as close to the problem as possible. With this, most policy areas are devolved to European member states, and only in areas where there is a clear need for distinct EU-level action does the EU have exclusive jurisdiction. As a consequence, such jurisdiction applies only to a handful of policy fields, such as the customs union,

[9] For an introduction, see Wallace, Pollack and Young (2014).

the common commercial policy and competition policy. In all other policy areas, member states share jurisdiction with the EU or act with its support. Energy, the environment and trans-European networks are shared competences.

The 2007 Lisbon Treaty, which forms the overarching legal framework for the EU's functioning, sets a high bar for EU-level action in areas that do not fall within the EU's exclusive competence. EU policy is justified only where a given policy objective cannot be 'sufficiently achieved' at national level, or where the EU level proves to be 'better' because of the scale or effects entailed in a given policy goal (European Communities 2007, Art. 5 (3)). Particularly in policy areas that national governments have traditionally regarded as strategic for national welfare or economic development, member states have proven very attenuative to securing and retaining their national prerogatives. This claim holds particularly true in the energy sector. National-level authority is hardwired into the Lisbon Treaty, which states that '[EU-level] measures shall not affect a Member State's right to determine the conditions for exploiting its energy resources, its choice between different energy sources and the general structure of its energy supply' (European Communities 2007, Art. 194 (2)). Even though energy represents a shared competence, Art. 194 therefore shields national prerogatives and gives member states a free hand in choosing their own energy mix. This, obviously, includes shale gas, to the extent that it is a nationally available resource.

The EU's actions in the realm of the environment and climate – both of which are relevant for conventional and unconventional energy production – are by and large based on Arts 191, 192 and 193 of the Lisbon Treaty. Art. 191 defines environmental protection and combating climate change as an EU policy goal. (Art. 192 essentially details procedures, while Art. 193 allows individual member states to adopt more progressive policies, should they wish to.) It sets the basis for important EU environmental action, for instance in the shape of the Directive on Environmental Impact Assessments (European Parliament and the Council 2014). The EU has also adopted ambitious climate targets, notably in the shape of the 20–20–20 goals, which call for 20 per cent renewables in the national energy mix, a 20 per cent increase in energy efficiency and a reduction of carbon emissions of 20 per cent by the year 2020. Among other things, a European Emissions Trading scheme was set in place, aimed at 'putting a price on carbon'

(European Commission 2009, 2014e; European Parliament and the Council 2009a, 2009b, 2009c).

Further, it is the declared goal of the EU to retain global leadership in climate action (Groen, Niemann and Oberthür 2012). Up to 2050, the EU's energy 'roadmap' therefore puts the bloc on a firm path towards a low-carbon economy, aiming to reduce greenhouse gas emissions by 80–95 per cent compared to 1990 levels (European Commission 2011a). With the EU moving towards an Energy Union, the Commission has tabled a 'Clean Energy for All Europeans' package, reinforcing the transition towards a low-carbon economy and aiming to generate new regulation in that regard (European Commission 2016b). In short, although Art. 194 secures member states' energy choices, environmental and climate policy set a broader framework in which for these to happen, notably through environmental provisions and their carbon impact.[10]

In fact, low-carbon policies have started to become a constraining factor in member states' energy policy choices. Climate action is a particularly contentious issue for countries relying heavily on fossil fuels in their power production – many of which are found in Eastern Europe. Poland's electricity supply, for instance, is almost 90 per cent dependent on coal. In the Czech Republic, coal covers more than 50 per cent of the national power supply, and even in gas-rich Romania, it makes up 26.9 per cent (Eurocoal 2017). This puts these countries at risk of running afoul of EU carbon policies going forward, with possible negative consequence for their economies and consumers. Even though the 'Golden Age of Gas' (as it was famously proclaimed by the IEA, 2012a) might not fully materialize on the EU level, gas is thus still a 'bridge fuel' for many Eastern European governments, which could help them transition to a low-carbon economy in the long run. For these countries, a key benefit of domestic shale production would therefore be the switch of their energy system from CO2-heavy coal to less polluting natural gas.

To date, the EU has not issued specific regulation or legal frameworks tailored to unconventional hydrocarbon extraction. Prompted by the European Parliament, the Commission initially sought to establish a defined legal framework for unconventional energy, dubbed the

[10] For a detailed legal discussion of EU environmental policy, see Jans and Vedder (2011).

'Shale Gas Directive'. Following interventions from the UK and Poland, Brussels eventually preferred leaving the regulation of unconventional hydrocarbons to member states (EUObserver 2012). Instead, the Commission, in 2014, issued a Communication flanked by a set of non-binding Recommendations on shale gas, pertaining to planning and licensing, environmental risk, well integrity, air emissions and best practice in operating sites (European Commission 2014b, 2014c).[11] According to the Recommendations, unconventional gas operations are, in principle, covered by pertinent EU frameworks. In essence, as long as EU member states abide by health and safety standards in the mineral-extracting industries, grant exploration licences and authorizations in a non-discriminatory manner, have companies manage waste water properly, monitor and report emissions, register chemicals as required and ensure environmental impact assessments are carried out, there is no need for specific EU action.

With this, it is the Commission's 2014 Recommendations coupled with (mostly) environmental legislation that provides the EU-level regulatory framework in shale gas. Key EU-level frameworks in the shape of Regulations and Directives comprise the Directive (and Regulation) on Registration, Evaluation, Authorization and Restriction of Chemical Substances (European Parliament and the Council 2006c, 2006d), the Groundwater Directive (European Parliament and the Council 2006b) amending the Water Framework Directive (European Parliament and the Council 2000), the Mining Waste Directive (European Parliament and the Council 2006a), the Strategic Environmental Assessment Directive (European Parliament and the Council 2001) along with the Directive on Environmental Impact Assessments (European Parliament and the Council 2012a), the Hydrocarbons Licensing Directive (European Parliament and the Council 1994), the Environmental Liability Directive (European Parliament and the Council 2004), the Industrial Emissions Directive (European Parliament and the Council 2010b), the Habitat Directive (European Council 1992a), the Oil and Gas Safety Directive (European Council 1992b), the COMAH Directive on Control of Major-Accident Hazards (European Parliament and the Council 2012b), the REACH Directive on

[11] As Reins (2017) notes, the Commission's opting for a mere Recommendation is consequential, given Art. 194 securing national choices on energy sources coupled with a strong subsidiary principle.

Dangerous Substances (European Parliament and the Council 2006c), the Offshore Oil and Gas Directive (European Parliament and the Council 2013) and pertinent provisions in the Birds Directive (European Parliament and the Council 2010a) and the Directive Ensuring Equal Access to Exploration Licences and Authorizations (European Parliament and the Council 1994).[12] As Boersma and Khodabakhsh (2014) note, however, pertinent EU-level legislation was put in place before unconventional gas emerged as a policy issue, and with it the fracking technique. While there exists a need to streamline the existing regulatory patchwork going forward, particularly regarding an unconventional gas industry, member states eager to push the technology might retain a strong interest in preventing Brussels from taking further action in this regard.

The 2015 move towards establishing the Energy Union, which was catalysed by the Ukraine crisis starting in 2014, adds a final dimension to European energy policy-making. Against the backdrop of increasingly tense Russia–EU relations, the Energy Union started off with a strong security spin, following Polish Prime Minister Tusk calling for an 'end of Russia's energy stranglehold' (Tusk 2014). Soon, however, debates moved towards integrating what are thus far somewhat distinct and scattered EU policies. This resulted in the Energy Union comprising five dimensions: energy security; energy market; energy efficiency; decarbonization; and research, innovation and competitiveness. Although observers noted that there exists a trend towards stronger securitization in EU energy policy (Boersma and Goldthau 2017; Judge and Maltby 2017) – epitomized by energy security's now topping the policy agenda – the Energy Union, at its core, represents an important step towards uniting energy, climate and environmental policies under one common institutional umbrella (Andersen, Goldthau and Sitter 2017).

For its lack of a clearly defined agenda beyond the rather general intention to help the EU 'meet its climate and energy policy goals until 2030' (European Commission 2017), the Energy Union has been criticized as remaining not much than 'an empty box' (Szulecki et al. 2016). Eventually, however, the Energy Union will align policies pertaining to the gas and power market, renewable energy, energy

[12] For a comprehensive list detailing pertinent Regulations and Directives, see Boersma and Khodabakhsh (2014) and European Commission (2014a).

efficiency and energy security. Its governance – as proposed in the Commission's draft regulation – will, among other things, require governments to produce integrated energy and climate plans, give the Commission a say in the latter and to a certain degree, therefore, streamline national policies further (European Commission 2016e). Still, the Energy Union is not susceptible to add a separate EU layer of policy-making to existing frameworks, do away with national competences in energy policy or specifically question member states' choices regarding shale gas.

2.5 The Road to Shale: An Eye on the 'Above Ground Factors'

The rise of the American unconventional gas industry, legend has it, is the story of a few Texan oilmen who, led by entrepreneurial spirit and strong determination, would eventually find the magic formula for extracting hydrocarbons from deep-seated shale rock formations (Gold 2014; Zuckerman 2013). George Mitchell, the legendary gas baron, is at the core of that story, which ends with fracking becoming the driver of the US 'shale revolution'. What is missing from this, however, is a few important enabling factors which allowed the pioneering 'frackers' to make hydraulic fracturing, a technology known since the 1940s, a commercially viable option. In other words, public policy mattered as much as entrepreneurship.

In fact, the regulatory context in the USA – on both the federal and the state level – was such that it favoured the development of shale gas, facilitated the diffusion of the fracking technology across the country and allowed the US shale gas industry to scale up (Rabe and Borick 2013; Wang and Krupnick 2013). This added to a competitive domestic market environment which fostered innovation and risk taking.

In Europe, many – if not most – of the enabling factors characterizing the US shale gas story are essentially absent (Boersma 2015; Boersma and Johnson 2012; Gény 2010; McGowan 2014; Stephenson 2016). Leaving aside geology – an aspect by and large pertaining to data availability in the European context – it is these 'above ground factors' (Goldthau 2016b) that Europe grapples with. Europe's struggle with defining an appropriate regulatory and institutional framework for shale gas therefore goes far beyond the EU being somewhat of a 'messy polity'. In fact, there exist fundamental differences compared to the USA, the 'motherland of fracking'.

A first, and much discussed element in this context is the ownership structure of subsoil resources. While the USA leaves mineral rights to landlords, Europe – and with it, the rest of the world – maintains a firm state hand on oil and gas reserves. The almost proverbial American farmer-turned-millionaire, thanks to shale gas production on his property, will therefore hardly be seen in the European context. Individual land owners also face little financial incentive to facilitate the prospecting and exploration of minerals on their property. To be sure, different mineral rights regimes do not prevent the development of unconventional gas. They certainly do, however, bring additional actors to the shale game, who are likely to represent at least pivot players, if not veto players. In order to co-opt local-level actors such as land owners and municipal communities, they will need to be given a material stake in natural gas extraction, which represents a call on the institutional design of revenue-sharing mechanisms across governance levels.

A second element pertains to energy market maturity. As already discussed, European gas markets are in the process of being integrated, but they have a long way still to go. Compared to the USA, the pan-European gas market is at an early stage still, and is populated by a small number of competitors. More to the point, while the USA is home to thousands of private mid-scale producers supplying a competitive and liquid, hub-based domestic market,[13] Europe not only sources the bulk of its gas demand from abroad, but its upstream segment is biased towards offshore production, not onshore E&P, and is dominated by just a few large multinational corporations, such as BP, Shell and Total, in addition to vertically integrated state corporations such as PGNiG. As a corollary, the deep and highly specialized service sector characterizing the USA is absent in Europe. Energy infrastructure, in turn, primarily connects non-EU gas reserves with European consumers, rather than facilitating intra-European gas distribution and trade. This compares to the US situation of highly integrated regional markets connected by a nationwide pipeline network and linked up to neighbouring countries. Commercializing European shale will therefore need to overcome the double challenge of having to operate in the context of low market maturity and a gas sector traditionally focusing on offshore production and long-distance trade.

[13] See Boersma (2016) for a 2016 snapshot.

A third element relates to the social acceptance of unconventional gas, and more specifically of the contested fracking technology. Public opinion on hydraulic fracturing differs across European countries. Generally, however, and with the notable exception of Poland, the attitude towards shale gas is rather negative (Lis et al. 2015). On one hand, this is a function of both a higher population density in Europe compared to the USA and ecological sensitivities among the population, coupled with strong environmental movements dating back to the 1970s and 80s. Pertinent issues discussed in the context of fracking, including groundwater safety, air pollution, methane leakage, wastewater disposal and disrupted wildlife (Jackson et al. 2014), have resonated negatively with environmental constituencies. On the other hand, European countries have little experience with an oil and gas industry being embedded in local contexts. To be sure, a long history in coal mining, for instance, has exposed many local communities, and indeed entire regions, to the extractive industries, leading to what transition scholars termed a 'co-evolution' of technology, social structures and political institutions (Geels 2005; Geels and Kemp 2007). In the oil and gas sector, however, such a co-evolutionary process did not emerge – a situation fundamentally different from that in the USA, where Texas, Pennsylvania and Ohio feature a longstanding history in highly 'localized' oil and gas production, and where local resistance has been much less pronounced (Brown et al. 2013; Evensen, Clarke and Stedman 2014). As a consequence, anti-fracking initiatives emerged across Europe, with some protests remaining local (such as in the UK and Poland) and some going national (such as in Bulgaria and France). Moreover, against the backdrop of mounting public concerns over the impact of fracking fluids on groundwater safety and the local environment, several countries imposed complete bans or de facto moratoria on unconventional gas production, including France, Bulgaria and Germany.[14] European shale therefore starts off in a comparably more demanding situation than that in the USA.

A fourth element is public policies supportive of nascent technologies. In Europe, fracking may find it hard to make it through the 'Valley of Death' that lies between the initial stage of technology development

[14] Note, however, that since 1950, fracking has been used for well stimulation in conventional oil and gas production in Europe. Germany, for instance, reports more than 300 recorded 'frack jobs' (Landesamt für Bergbau, Energie und Geologie 2013).

and full commercialization (LaBelle and Goldthau 2014a). While the fracking technique clearly is mature in the USA, it will need to prove competitive in market environments beyond there. The Oxford Institute for Energy Studies puts production costs for European shale at USD 8–12 per MMBtu (Gény 2010, 87), and a study by the EU Commission's Joint Research Centre suggests costs of between USD 5 and 12 per MMBtu (Pearson et al. 2012, 160). For the UK, Bloomberg New Energy Finance estimated a cost range of USD 7.10 to 12.20 per MMBtu (House of Lords 2013). While these preliminary estimates need to be treated with caution, break-even prices for European shale seem to be significantly higher compared to the USA, where extraction costs are reported to be at USD 4 per MMBtu for the bulk of plays (IHS Energy 2016). More importantly, a European gas price environment of USD 8 per MMBtu or less[15] might present a significant challenge for the commercialization of shale gas. One way in which the USA addressed this problem was through R&D programmes and tax credits, which helped private companies make precisely the investments that led to the large-scale development of unconventional natural gas, even in a difficult price environment, thereby bringing down extraction costs (Wang and Krupnick 2013). Targeted policy measures such as the 2005 exemption of fracking fluids from the US Clean Drinking Water Act (famously dubbed the 'Halliburton Loophole', due to then-Vice President Dick Cheney, the former Halliburton CEO, throwing his weight behind the exemption) effectively functioned as a subsidy for the nascent shale industry.

Technological progress might, over time, improve the economics of European shale. Yet, as the Commission warns, the learning curve may not be as steep as in the USA. Instead, European shale extraction will likely come with a significant and persistent cost premium of 50 per cent on capital costs and 25 per cent on operating and maintenance costs (European Commission 2014j). Clearly, it would be unrealistic to expect European governments to support shale development with similar fiscal policies – let alone adopt steps similar to the Halliburton Loophole – not the least against the backdrop of adverse public opinion. But countries keen to develop an unconventional gas industry, notably in Eastern Europe, will need to put in place public

[15] This figure represents the 2015 all-year average (IGU 2016).

policies encouraging risk-taking in E&P and eventually improving the economics of shale gas extraction.

Finally, in Eastern Europe, the post-Communist legacy discussed earlier forms an important above ground factor in shale gas. Incumbent corporate stakeholders tied to traditional LTC contracts, regulatory regimes favouring state-run corporations and an energy system still by and large mirroring the public utility model characterize large parts of CEE. This legacy brings about a bias in the state apparatus, and indeed the national system of energy governance. These specifically regional characteristics add to the broader European intricacies in shale, and generate a specific regional context. The additional challenge therefore consists in overcoming incumbent material interests, institutional legacies and deep-seated structural patterns in regulatory governance.

As the subsequent chapters will demonstrate in more detail, it is national policy regimes that will determine how these above ground factors shake out; that is, whether social acceptance is addressed, incumbent companies (and socio-economic stakeholders more generally) are brought in, regulatory regimes are adapted and local-level communities see a material benefit of shale gas development. The way ideas, institutions and interests are aligned may or may not generate precisely a 'social licence to extract' shale gas in Eastern Europe. With this, we turn to a detailed discussion of the analytical framework underpinning this book.

3 *The Analytical Context*

Policy Regimes and the Social Licence

3.1 Regimes in Policy Analysis: A Brief Review of the Concept

As this chapter argues, the concept of policy regimes offers a powerful tool for explaining divergent policy trajectories in Eastern European shale gas. More to the point, it outlines that it is the specific interplay of ideas, interests and institutions that generates divergent policy outcomes on national levels, and hence allows the complex dynamics in domestic shale gas politics to be disentangled. The regime concept has a longstanding career in political science inquiry, and features in a diverse set of disciplinary subfields. Unsurprisingly, each of these subfields tends to use the regime concept in very distinct ways and puts it to work differently. Before delving further into the details of the analytical framework of this book and its operationalization in terms of comparative public policy, the notion of policy regimes therefore warrants some discussion in the broader context of social science research. In what follows, we briefly review key aspects pertaining to the concept of regimes, and on four levels of analysis: international (IR), supranational (EU studies), national (policy studies) and local (urban politics).

In the field of International Relations, the regime concept is used to study cooperation and conflict. It had particular traction in the liberal strand of IR thought. Regimes primarily constitute an option to overcome coordination problems among states and establish international social order under conditions of anarchy. As famously defined by Krasner (1983), regimes consist of 'principles, norms, rules and decision-making procedures around which actors expectations converge in a given issue area' (2). This definition points to the ideational element, institutional structures and processes that form the core of regimes. The basic function of regimes is 'to coordinate state behaviour to achieve desired outcomes in particular issue areas' (6), by way of providing information and enhancing compliance among involved actors (Keohane 1984). Regimes are considered crucial for facilitating coordination in key policy areas such as trade, finance and health, and they constitute a determinant of the strength of international regulation (Drezner 2007). While (state) actors give rise to regimes and are crucial for maintaining them, regimes – in the shape of international institutions – may also shape actor preferences and behaviour (Wendt 1994).

The policy literature generally subscribes to the broad notion of regimes as regulatory frameworks, albeit with a different analytical focus. Analyses typically investigate the effects of regime changes on administrative structures (Harris and Milkis 1996), institutional patterns and processes (Eisner 1994), national policy change (Jordana, Levi-Faur and Puig 2006) or regulatory divergence (Hood, Rothstein and Baldwin 2001) over time (Newman and Howlett 2014). Although not restricted to the national level, investigations typically retain an interest in federal- (or EU) and sub-federal-level dynamics pertaining to regimes – both as a cause of policy and as its effect. A large literature on policy subsystems adds the idea that policy-makers formulate and implement policies jointly and as part of a constant feedback process with stakeholders in society (Jordan 1990; Sabatier 1988; Stein and Bickers 1995). Depending on the issue at stake, the policy domain and the nature of the institutional environment, regimes and regime dynamics may differ considerably (Baumgartner and Jones 1991; Jenkins-Smith, St Clair and Woods 1991; McCool 1998).

Operating somewhat between the international and the domestic level, the Europeanization literature has embraced and adopted the regime concept. It is particularly against the backdrop of the EU being characterized as a regulatory state (Lodge 2008; Majone 1997; McGowan and Wallace 1996) that regimes have become central to analysing European policy-making and policy change. Pertinent studies, for instance, investigate the way EU regulatory regimes impact on national-level policies, how national administrative structures change in the face of EU regulatory pressure and the reasons for divergence in national-level responses to the EU regime (Knill 1998; Knill and Lehmkuhl 2002; Levi-Faur 1999; Lodge 2002). In this context, a more recent debate has come to centre on the various modes of governance the EU has experimented with in different policy areas, ranging from 'soft' (Open Method of Coordination) to 'tough' (e.g. surveillance mechanisms in fiscal policy) (Kohler-Koch and Rittberger 2006; Sabel and Zeitlin 2010).

Adding a local-level perspectives, urban studies embrace the regime concept when asking how, 'in a world of limited and dispersed authority . . . actors work together across institutional lines to produce a capacity to govern and to bring about publicly significant results' (Stone 1989a, 8–9, cited in Lauria 1996, 4). The conceptual distinction between state and market actors in urban regimes has led to long-standing debates (Imbroscio 1998), a key feature of how and to what extent regimes bridge the public–private divide (Elkin 1987; see also Dowding 2001). Although the IR and the public policy literature have also come to recognize the important role played by non-state actors in formulating and implementing policy, it is probably the urban studies literature that has most focused on how private and public interests can combine to foster the public good (Davies 2002; Mossberger and Stoker 2001).

Drawing from these strands of literature, policy regimes have been framed as 'governing arrangements for addressing policy problems' (May and Jochim 2013, 429). In essence, policy regimes determine which social, state and market players interact in the policy process, and how this interaction is structured (Eisner 2000). With this, the notion of policy regimes subscribes to key elements of Krasner's definition, but remains open as regards the types of actors forming part of the regime and the institutional design structuring their interactions. Also, in the policy literature, the concept of policy regimes has informed very diverse sets of scholarly inquiry in public policy. For Howlett (2009), for instance, policy regimes are part of a staged process of policy instrument choices and policy design; Orren and Skowronek (1998) use the term for analysing altering constitutional relationships in the USA; and for McGuinn (2006), a policy regime defines the mechanisms of policy change.

What's more, 'policy regime' is also used differently across policy areas. In environmental policy, it delineates paradigms underpinning policy-making (Jahn 1998) or separates novel governance arrangements from traditional types of regulation (Howlett and Rayner 2006). In social policy, it denotes different types of welfare regimes (Esping-Anderson 1990). Scholars of monetary policy refer to policy regimes as institutional arrangements coupled with a set of (policy) expectations (Bordo and Schwartz 1997).

Finally, there does not exist any consensus on how policy regimes emerge or change. Causes prominently referred to in the literature include 'external stressors' and 'shocks' (Wilson 2000), economic crises (Williams 2009) and feedback loops (Weaver 2010).

As this brief review reveals, a more detailed elaboration of the concept is warranted. The next section develops further the notion of policy regimes and sketches the analytical framework for this book.

3.2 Interests, Institutions and Ideas: Defining the Analytical Framework

For all their differences, existing academic works agree on the general notion that policy regimes consist of a policy paradigm, a power arrangement and an organizational arrangement (Wilson 2000) – in essence, a 'set of ideas, interests, and institutions that structures governmental activity in a particular issue area' (McGuinn 2006, 206). Turning to interests first, these essentially represent societal, economic or political stakeholders in a given issue area who may face gains or losses from a new policy or from policy change. In political economy terms, 'interests' may constitute the necessary support for a policy if there is material opportunity or benefit. Incumbents, in turn, may represent vested interests that could give rise of societal resistance against a policy. To be sure, winners and losers exist almost whenever policy change occurs. The ones that matter, however, are key stakeholders – societal actors who are crucially needed to ensure the success of a policy (Atkinson and Coleman 1989), or who may emerge to veto players in the policy process because they have the means and influence to mobilize against them (Carmine, Darnall and Mil-Homens 2003). In short, the 'power arrangement' determines which stakeholders are part of the regime, are given access to the policy process and may shape this process.

In addition to representing material motives, it is also important to recognize interests as 'constituencies' (May and Jochim 2013, 434). As constituencies, interests not only provide support and organize opposition to a given policy but also determine the governing capacity of a regime (May and Jochim 2013, 434). This is because the regime concept – in line with key premises of the governance literature – acknowledges that single (state) actors can only rarely coerce others into an arrangement that lasts. The essential idea here is that elites can barely exert control over a policy area or 'deliver' on policy goals on their own, and that policies are therefore brought about not by hierarchical state control but in conjunction with and through the participation of non-state actors (Howlett and Ramesh 2002; Rhodes 1997). This point is underlined by the trend towards devolution, in which governments 'produce' public goods and services in conjunction with for-profit and non-profit organizations (Kettl 2000). This is both a deliberate choice and a necessity, given the complex settings in which modern societies operate. As a consequence, state agencies crucially rely on societal actors and businesses to formulate policy and implement it (Milward and Provan 2000). Research empirically demonstrates the importance of interests as 'constituencies' for policy areas as different as river basin and flood risk management, in which local organizations participate in the planning process (Watson, Deeming and Treffny 2009); urban planning, where entrepreneurs, associations and voluntary sector organizations represent crucial stakeholders neighbourhood development (Kazepov 2004); and education policy, where 'boundary-spanning leadership' (Burr 2005) seeks to tie communities and social and private actors back to school reform processes and give them a say (Horsford 2010). In the context of the fragmented power characterizing modern society and policy-making, regimes therefore emerge as 'the collaborative arrangements through which . . . governments and private actors assemble the capacity to govern' (Mossberger and Stoker 2001, 812).

An important aspect in this context is legitimacy, a concept resting on the key categories of justness and appropriateness (Dahl 1998). The aspect of legitimacy is more than a matter of moral philosophy. Rather, 'bringing interests in' is viewed as important to enhancing the democratic accountability of a given policy, as well as its effectiveness, efficiency and performance. Delivering results increases the output legitimacy of policies 'because they effectively promote the common welfare of the constituency in question' (Scharpf 1999, 6). This, clearly, is an argument pertaining to material motivations. At the same time, the process of stakeholder involvement may also foster input legitimacy and strengthen the acceptance a policy enjoys even before it yields tangible results (Crozier 2010; Rothstein and Teorell 2008; Scharpf 1999). Here, the focus is more on democratic processes, representation and accountability. While there is an ongoing debate about the extent to which the focus has gradually shifted from input to output – or even 'throughput' – legitimacy (Peters 2010; Schmidt 2013), the key point is that private and social actors are legitimate – and, in fact, crucial –

players in the policy process. To be sure, participatory governance per se is not a panacea for contested policy issues (such as the ones surrounding risky technologies), and accounting for stakeholders on various policy levels may indeed create its own 'tyranny' (Cooke and Kothari 2001; Hickey and Mohan 2004). Yet, the most relevant insight for this book is that interest representation is key both for reasons related to material motives and to ensuring a given policy is regarded as just and appropriate among the affected constituencies. Juxtaposed with a political economy lens on interests, the different conceptualization becomes clear: a political economy take would look at interests in terms of how to empower winners and compensate losers in order to deal with policy contestation, while a policy regime perspective would stress the due process of interest inclusion, with a view to committing societal and business stakeholders to a commonly shared policy goal.

Institutions, the second element of the policy regime framework, comprise what Krasner (1983) subsumes under the rubric of decision-making procedures, norms and rules. Institutions principally provide incentives and constraints for state and non-state players and shape the ways they operate (North 1990). A more encompassing definition of institutions gives room to administrative structures, state institutions and their organizational relationship (Hall and Taylor 1996). In policy regimes, institutions are relevant in a number of ways. Most importantly, they define the procedures and processes pertaining to decision-making and policy implementation. That way, the institutional set-up – or, in Wilson's (2000) terms, the 'organizational arrangement' – 'structures authority, attention, information flows, and relationships' (Jochim and May 2010, 313). In essence, therefore, institutions define who has a say and who doesn't, shape the mode of interest representation (inclusive or exclusive) and its quality (top down or bottom up) and distribute policy ownership among the involved state bureaucracies and non-state actors. As a corollary, and depending on which stakeholders are included, the institutional structure may therefore lend authority to the governance arrangement and in that way give legitimacy to a policy agenda and its underlying goals. In this context, the impact of institutional arrangements on information flows deserves specific mention. As will be argued in more detail when we delve into Eastern European shale gas, information represents both a means of including (or excluding) key constituencies and a way of empowering stakeholders that have no formal role to play in the institutional set-up.

A good example to illustrate this point is the Dutch approach to managing the Netherlands' energy transition to a low-carbon future. This approach was characterized by a comprehensive process aimed at including societal stakeholders and deliberating on both the low-carbon pathways and the ways to achieve them. Thus, non-established players were given policy ownership, while incumbent actors such as companies and regulators were checked, assigned a stake in the process and given a role in planning and implementation, in addition to being organizationally tied back to state agencies (Kemp, Rotmans and Loorbach 2007). At the same time, the expansion of Dutch wind power capacity suffered from serious setbacks during its early phases, which besides minor not-in-my-backyard (NIMBY) problems was due to flaws in the institutional set-up governing siting procedures, notably related to a top-down policy style and institutional inertia (Wolsink 2000). Institutional procedures were therefore found to be as important as inclusive power arrangements.

Further, institutions and institutional design also entail an element of political control. As McCubbins, Noll and Weingast (1989) show for the case of the US Environmental Protection Agency (EPA), politicians deliberately use institutional structures and processes to monitor and influence the behaviour of bureaucracies. Moreover, legislators may purposefully design administrative structures and processes in such a way that regulatory agencies generate the desired policy outcomes. Strategies include defining the mission of the agency in narrow terms and restricting its powers to specific policy areas, as a result of which regulatory decisions cater to the legislator's political preferences (McCubbins, Noll and Weingast 1987). The Texas Railroad Commission (TRC), for instance, was assigned the task of regulating the Texan oil and gas industry, albeit primarily with a view to promoting the state's (fossil) energy sector (Rahm 2011) rather than environmental protection – a mission it arguably lived up to fairly well. An example of limiting the regulatory scope of an agency is presented by the US 2005 Energy Policy Act, in which fracking was exempt from the US Clean Drinking Water Act, and hence from EPA ruling. The Bush administration's obvious purpose here was to foster the nascent shale gas industry.[1]

[1] For a more detailed discussion on the regulatory governance agenda in shale gas, see Goldthau (2016b).

Finally, institutional quality has been broadly correlated with policy effectiveness. The main finding from a broad scope of empirical studies is that there exists a correlation between the quality of the public service and the institutional procedures governing it, and the quality of policy implementation. For instance, Meier and Smith (1994) link the failure of the Drug Enforcement Agency to effectively administer US drug laws to flaws hardwired into the bureaucracy's set-up and the way it was designed to implement these laws. Moreover, a robust link has been established between 'good governance' and the support among the population and stakeholders for bureaucratic decisions (Anderson and Tverdova 2003). This extends to a country's economic performance, which is found as correlating with institutional quality (North 1990). Empirical investigations of developing countries' efforts to increase household income support the relative importance of institutional quality over other factors, such as geography (Rodrik, Subramanian and Trebbi 2004). They also seem to explain the success or failure of foreign aid (Dollar and Levin 2005). As this research suggests, the quality of bureaucratic structures, legal frameworks and administrative processes not only determines whether policies are implemented well and deliver results, but also affects the transaction costs that private actors in particular face when operating.

Ideas, the third element, refer back to Krasner's notion of 'principles', defined as 'beliefs of fact and causation' (Krasner 1983, 2). Ideas have played an important role in the analysis of public policy and regulatory politics. For instance, as Harris and Milkis (1996) show, the US push for the deregulation, privatization and liberalization of many publicly run sectors in the 1980s is not primarily a function of the Reagan administration seeking to enhance economic growth and end recession; instead, it can be attributed to 'weightier questions of democracy, citizenship, the evolution of an administrative state, and the role of ideas in American politics' (3). As Goldthau (2012b) argues for the case of the energy industry, governance models in the oil and gas sector also swing, in terms of policy ideas (or paradigms), from the state to the market and back. In other words, it is beliefs that inform and shape policy. Further, in his study on race policy in Britain and France, Bleich (2002) identifies ideas – or frames – as a determinant for explaining cross-country policy differences. Frames allow actors to generate interpretations of a given policy problem, but also enable them to define and possibly constrain the range of policy options available to address that problem. Moreover, as Baumgartner and Jones (1991) argue, the interaction of beliefs and values concerning a particular policy – the policy image – can also be causal for the way policy evolves. Their studies show that a policy image turning negative has been a decisive factor in reversing the USA's policy towards nuclear power.

The concept of frames travels even further. Research on technology policy – an important branch of work for this book – suggests that ideas are decisive for shaping how society and technology intersect. Here, scholars of the social construction of technology have coined the term 'interpretative frames' to refer to the meaning that is attributed to a novel technology (or 'artifact') by relevant social groups (Bijker 1997; Law 1991; Pinch and Bijker 1984). In the words of Klein and Kleinman (2002), technology acquires meaning through a 'shared cognitive frame that defines a relevant social group and constitutes members' common interpretation of an artifact' (31). The key point here is that it is not necessarily the detailed technical characteristics – risky or safe, large-scale or small-scale, disruptive or consequential – that are decisive for the socio-economic acceptance, and hence success, of a novel technology. Rather, it is its interpretation by relevant societal groups, i.e. 'what they make of it'. Indeed, shale gas – more precisely, fracking – is a prime case for a novel technology. The way the new technology is framed among societal constituencies also defines the focus of pertinent debates in the policy process. As an emerging set of research shows, such frames differ widely across Europe, and often even among societal groups and constituencies within countries (Goldthau and Sovacool 2016; Williams et al. 2015).

While political ideas can be thought of as 'free-floating bits of knowledge and conjecture, detached from considerations of structure and power' (Liebermann 2002, 700), they can also be turned into policy narratives, with a view to strategically defining a problem. In this case, ideas serve as 'narrative story lines and symbolic devices to manipulate . . . issue characteristics' (Stone 1989b, 282), so that an individually preferred policy solution resonates with the broader public (Stone 2002). Yet, ideas are about more than giving a policy problem a certain spin, in order to ensure support or organize opposition. 'Core beliefs' as shared by a critical group of people are considered central to forming advocacy coalitions for a certain policy (Sabatier and Jenkins-Smith 1999). In this context, it is particularly policy entrepreneurs who may use ideas as 'tools of coalition building' (Eisner 1994, 158) in order to champion a policy or foster change. In fact, policy entrepreneurs are key to building momentum for an idea in the policy process. As Kingdon (1984) pointedly puts it, '[g]ood ideas lie fallow for lack of an advocate'. Ideas may, however, also serve incumbents. A case in point is what Baumgartner and Jones (1993) refer to as 'policy monopolies', a situation in which powerful ideas, in conjunction with 'definable institutional structures', enable policy entrepreneurs to retain a favourable status quo (7).

Against this backdrop, May and Jochim (2013) conceptualize ideas as the 'glue' of a policy regime (435). In their framework, ideas serve as the political narrative that gives meaning to a policy and generates support for it among stakeholders. In order to specify the latter point further, the works of Cox and Béland (2012) are informative for the purpose of this study. As they argue, a specific idea must be of 'high valence' to gain traction; that is, the portrayed image of a policy has to be attractive and engaging. As Cox and Béland show for the case of energy security and environmental sustainability, an idea of high valence allows policy entrepreneurs to frame a policy issue in such a way that it resonates with core constituencies, generates support among key stakeholders and therefore likely influences policy change (308).

In sum, and slightly refining McGuinn's (2006) definition, this book conceptualizes policy regimes as *a governing arrangement consisting of ideas, interests and institutions that generates policy in a particular issue area*. This definition is inclusive (comprising state and non-state actors), acknowledges institutional structures and power, and provides room for ideational factors to exert distinct and independent influence. With this, the policy regime concept allows the conceptualization of the above ground factors of shale gas in a single framework. More to the point, it serves three purposes for this book.

First, it presents a heuristics device for exploring the comparative public policy of shale gas in Eastern Europe. Each of the country cases studied in this book features distinct actor constellations, institutional structures and regulatory processes pertaining to shale gas, in addition to policy narratives as used by key political or societal actors to spin the debate. The policy regime framework allows these complex domestic dynamics to be unpacked by giving due attention to the role played by state and non-state actors in the policy process and the institutional and ideational environment shaping their interaction. It represents a means for disentangling the specific characteristics of each country's 'governing arrangement' by way of 'backward mapping' these characteristics (May and Jochim 2013, 427). This allows for thick descriptions of national contexts, accounting also for inconsistencies and nonlinearities in domestic policy processes in shale. With this, the book essentially takes to heart Lieberman's (2002) claim that 'analysis that takes both ideas and institutions seriously will almost of necessity shed light on points of friction, irregularities, and discontinuities that drive political change' (698).

Second, policy regimes allow for an explanation of the success or failure of governmental policy agendas.[2] More specifically, the policy regime framework helps shed light on policy adoption and, to a certain extent, its implementation. To be sure, this analysis will refer to various stages of the policy process pertaining to Eastern European shale gas. Yet, it is primarily the question why shale gas policy agendas get 'stuck' in some countries and not others, and to what extent these agendas got reshaped before being adopted and implemented, that stands at the centre of analysis. As this study shows, shale gas policies have become subject to fierce political controversy, particularly since being put forward by various governments in Eastern Europe, with a diverse set of economic and societal actors joining in that controversy. Moreover, as the analysis demonstrates, the process of stakeholder involvement, in conjunction with specific institutional and procedural choices, represents a decisive element in shaping the outcome of these controversies and leads to highly divergent country-level policies. In that sense, this book acknowledges the rich literature on policy adoption and implementation (Elmore 1978; Howlett, Ramesh and Perl 2009; Lipsky 1980; Mazmanian and Sabatier 1989; Rein 1983; Sabatier 1986). Its main contribution to the literature, however, lies in exploring 'how policy regimes can serve as the political and institutional means for securing policy legitimacy, coherence, and durability' during the initial stages of the policy process stages (May and Jochim 2013, 443).

[2] Note that the terms 'success' and 'failure' solely pertain to whether a government's policy initiative is eventually adopted or not, and thus moves to implementation. In other words: this book clearly does not aim to champion a position on shale gas. Instead, it aims at explaining cross-country policy divergence in shale gas.

Third, the policy regime concept allows for the study of policy-making in messy contexts. In fact, Eastern European shale gas is a show case of policy-making in a fragmented setting: in regulatory terms, authority on key aspects such as licensing exploration activities, vetting environmental impact assessments and monitoring operations is spread across various governance levels, from the national to the local; as a policy issue, shale gas spans established and clearly delineated areas of policy-making, including environmental and human safety, economic development, industrial competitiveness and even national security; and as a sector, shale gas comprises constituencies from the public to the private sphere, including ministries of energy and environment, regulatory agencies, regional administrations in charge of environmental oversight, local municipalities responsible for infrastructure, state-owned and private energy companies active in natural gas upstream or distribution, or both, and environmental activists and other civil society groups. In addition, shale gas features characteristics of 'wicked' policy problems: it is strongly contested; solutions are not clear cut, but instead feature clear trade-offs; and it is characterized by contradictory certitudes among stakeholders (Head 2008; Rittel and Webber 1973; Verweij et al. 2006). As May et al. (2011) suggest, policy regimes may be conducive to bridging actor groups and governance levels, if designed with a view to injecting cohesive dynamics into a disjointed policy setting. Eastern European shale gas, therefore, presents a promising case to test this argument and to explore in more detail whether and to what extent policy regimes help overcome policy fragmentation.

3.3 Operationalizing Policy Regimes: Regime Strength and Shale Gas Policy

The main assertion made in this book is that policy regimes matter for the 'fate' of a government's shale gas agenda. More to the point, it is a policy regime's strength that is decisive for determining whether a government's shale gas policy agenda is adopted and becomes enshrined in pertinent regulatory frameworks, and whether the latter enjoy the legitimacy to be implemented. With this assertion, this book follows May et al. (2011) when it comes to operationalizing the policy regime concept for investigating Eastern European shale. May et al. (2011) define regime strength as 'the ability . . . to focus attention of [relevant] players on a shared vision . . . in addressing a given boundary-spanning problem' (290). Strong regimes generate cohesion in that context by way of overcoming fragmentation among actors, interests and policy levels. Weak regimes, by contrast, are likely to cause policies to fail as they leave stakeholders fragmented and interests scattered across governance levels. In other words, strong regimes align actors, institutions and policy narratives, foster a shared vision among key stakeholders, engage the latter and are characterized by institutional arrangements that support both the joint vision on a policy and its implementation. Weak regimes, by contrast, fail to do so.

As a corollary, this study subscribes to an elitist notion of agenda setting, i.e. it puts emphasis on the role of political decision-makers in initiating a policy (Kingdon 1984). In turn, and in line with the policy regime concept, societal actors and businesses are considered crucial when it comes to making this policy agenda reality. This conceptual choice heeds the call of Capano, Howlett and Ramesh (2015) to 'bring governments back in', stressing their pivotal role in setting up new governance arrangements (see also Atkinson and Coleman 1989; Howlett 2004). At the same time, it is informed by the empirical fact that it is governmental representatives who early on promoted shale gas development in Eastern Europe. This does not discard the role of societal actors or the media in formulating and promoting policy alternatives, but it puts the conceptual and empirical focus on the (divergent) trajectories of national shale gas policy agendas and the dynamics evolving around them.

Note that much of the existing literature on policy regimes focuses on the degree to which regime dynamics may induce policy change. In this context, policy subsystems (Atkinson and Coleman 1992; Freeman and Stevens 1987; Stein and Bickers 1995; Zafonte and Sabatier 1998) emerge as a key reference point, and it has been argued that policy regimes help span policy subsystem boundaries (May, Jochim and Sapotichne 2011) and support the establishment of new policy equilibria (Worsham and Stores 2012). This book, however, is not primarily interested in policy subsystem stability or change. Instead, its focus of analysis is on policy adoption and – in part – implementation, and the politics thereof. On one hand, this analytical choice is supported by the fact that much of the pertinent literature on policy subsystems was developed for a US context, including its system of parliamentary committees and subcommittees, legislative procedures and policy venues such as Congressional chambers, US courts and the White House. Arguably, this literature therefore is not entirely applicable to an Eastern European policy environment. On the other hand, the probably analytically more important point is that this book is primarily interested in exploring the degree to which regimes help explain policy divergence across countries. Here, the policy regime concept serves as an analytical device for investigating fragmentation and cohesion among actors, interests and policy levels. With this, the book does not disregard the rich existing literature on policy subsystems, but it has a slightly different scope: rather than aiming at disentangling complex policy subsystems and their dynamics, it is limited to exploring whether regime strength can be identified as a factor determining the success or failure of governmental policy agendas and how regime constellations of actors, institutions and ideas relate to policy outputs in Eastern European shale gas.

As outlined in Chapter 2, Eastern European governments championed shale gas across the board. A strong policy regime can therefore be assumed to mobilize key societal constituencies in support of the commonly shared policy goal of developing a national unconventional gas sector; to make these constituencies endorse a strong policy narrative surrounding the positive effects of shale for the nation's economic prosperity or national security; and to be grounded in institutional frameworks that facilitate the buy-in of relevant stakeholders, as well as policy coordination between and across governance levels. In that way, a strong policy regime can also be assumed to co-opt vested economic interests, for instance in the domestic energy sector, or even social movements, thereby decreasing societal resistance and lending legitimacy to shale gas policies as advocated by the government. A weak policy regime, by contrast, does co-opt crucial stakeholders, lacks an inclusive policy idea and fails to facilitate 'collaborative governance' (May and Jochim 2013, 435) across policy levels. From this, the book's general hypothesis follows:

GH: If the policy regime surrounding shale gas is strong, governmental policy agendas are adopted and implemented successfully.

Turning to the individual elements of policy regimes, interest involvement is operationalized in two ways: first, we identify relevant actors in the shale gas sector, for each country and governance level, based on their material, procedural or normative motivations; and second, we assess their empirical involvement in the shale gas policy process in each studied country. The assessment focuses on and measures the extent to which key stakeholders and constituencies in society and the economy lent support and momentum to the government's (pro-)shale policy agendas. The specific hypothesis reads as follows:

SH1: A comprehensive involvement of economic and societal interests fosters support for governmental shale gas policy agendas and facilitates their adoption, and eventual implementation.

In terms of institutions, we operationalize strength by way of defining key qualities that strong institutional arrangements bring to a regime. In essence, a strong regime features institutional structures and processes that are designed to address key aspects of 'wicked' problems such as shale gas policy, facilitate cooperation among the various stakeholders and enhance information flow among those stakeholders (Pollitt 2003). On the one hand, this has long been discussed under the rubric of 'joined-up government' (Davies 2009; Pollitt 2003), 'horizontal government' (Peters 1998) or 'whole-of-government' (Christensen and Lægreid 2007). Originating in the Anglo-Saxon world, the main goal of such policy initiatives was to make administrative bureaucracies deliver outcomes regardless of departmental boundaries, among other things by coordinating policies between different units and communicating effectively (Kavanagh and Richards 2001). This necessitated finding institutional structures and processes that help overcome bureaucratic departmentalism and enable 'joined-up *implementation*' of policy across administrative units. With this, the main concern moved away from following institutional logics (the administrative process) to concentrating on results (the output and outcome). In short, a holistic approach to delivering public policy was meant 'to overcome the limitations on policy effectiveness caused by layering, drift and conversion' (Howlett and Rayner 2006, 170). With regard to the state apparatus, a strong policy regime is therefore characterized by institutional arrangements that 'channel . . . attention, information, and organizational relationships in support of policy goals' and 'establish . . . meaningful linkages among relevant implementing authorities' (May and Jochim 2013, 436).

On the other hand, a strong policy regime would include 'participatory institutions' (Avritzer 2009) designed to reach out to societal actors with a view to including them in 'state-sanctioned policy making venues' (Wampler 2007, 256). While participatory institutions play an important role in the literature on policy formulation and decision-making processes, they are also important for rendering implementation processes more transparent, legitimate and, hence, effective. Inclusive institutional structures enable interactions between state and non-state actors, facilitate mutual information flow between citizens and government officials and, hence, resume an important function regarding communication and feedback (Baumgartner and Jones 2002; Lowi 1964; Schattschneider 1935). This gives us a second specific hypothesis:

SH2: Institutional structures providing venues for stakeholder participation, ensuring mutual communication and facilitating multilevel policy coordination foster support for governmental shale gas policies, and help their adoption and implementation.

We operationalize institutional strength as the degree to which: non-state actors from society and business can become part of a communicative process on shale gas agendas; relevant state authorities on all policy levels coordinate on shale gas policy and are given policy ownership; and the institutional arrangement demonstrates flexibility and allows for adjustments in procedures and processes.

Note that this operationalization of institutional strength also comprises what Howlett and Rayner (2006) would characterize as policy *instruments*; that is, formal or informal procedures for facilitating interest group involvement, mediation and arbitration. However, regulatory processes and related platforms – town hall meetings, public hearings, written petitions, etc. – not only serve as mere instruments for policy design, but, in the context of contested policy issues that require comprehensive stakeholder involvement, also constitute crucial elements of the institutional arrangements linking various governance levels and facilitating policy implementation.

Ideas, finally, are operationalized in terms of their 'emotional quality' (Cox and Béland 2012); that is, their 'valence', determining their attractiveness. Ideas with high valence resonate with various constituencies and foster a shared understanding of the goals and meanings of a policy. Ideas with low valence, by contrast, are not likely be shared by key stakeholders and generate opposition to a policy. In this context, it is important to bear in mind that novel technologies such as fracking are contested within society. Yet, from a social constructivist lens, this contestation is not necessarily about whether the fracking technology is 'safe' or comes with hazardous environmental side effects. Rather, it is about different interpretative frames that compete in an ongoing process of attributon of meaning to the novel fracking technology. To be sure, stakeholder support for shale, and their endorsement of a given policy narrative on fracking, may be driven by material interests. Yet, arguably, such material motivations alone can hardly form the core of a (policy) narrative on a contested technology. Instead, a given policy narrative on fracking will only resonate more broadly within society if it coincides with the frame that relevant groups attach to that technology. With this in mind, we have a third specific hypothesis:

SH3: Shale gas policy agendas succeed if underpinned by ideas of 'high valence' resonating with dominant interpretative frames among societal constituencies and ensuring stakeholder support.

Table 3.1 ***Policy regimes in Eastern European shale gas: hypothesis and indicators***

	Analytical focus	Key indicators (high/low)
Ideas	Valence of policy frames	• Degree of 'ideational uptake' among stakeholders • Congruence of dominant interpretative frames of governmental actors and societal groups
Interests	Constituency representation	• Involvement of stakeholders in policy process • Inclusion of veto players in policy process
Institutions	Communication	• Procedural outreach towards non-state actors • Transparency of policy process • Information flow among stakeholders
	Joined-up government	• Cooperation of relevant state authorities on all policy levels • Policy ownership across policy levels • Flexibility towards adjusting procedures and processes • Administrative quality

We operationalize valence as the extent to which relevant players from different societal groups and across various governance levels refer to similar interpretative frames when giving meaning to shale gas and fracking. In this context, it is the degree of 'ideational uptake' (May, Jochim and Sapotichne 2011, 292) among key stakeholders that matters. Table 3.1 summarizes the foregoing.

In sum, this book views policy regimes as governing arrangements that are constituted by societal interests, institutional structures and ideational frames. A strong policy regime mobilizes the support of core societal constituencies; ensures their endorsement of a common policy narrative; and facilitates policy coordination between and across governance levels.

3.4 The 'Social Licence to Frack'

Policy regimes not only help explain policy divergence across countries but, at least for the case of shale gas, can also create a 'social licence'. As discussed, fracking represents a highly contested extractive technique. Pro-shale policy agendas therefore warrant broad support among societal constituencies to eventually manifest in the shape of legal frameworks, and so too does their implementation. Strong policy regimes are susceptible to generating this societal support by way of aligning actors, institutions and policy narratives around the policy agenda and the policy process underpinning it. It is important that this assertion does not necessarily constitute a hypothesis of its own. Rather, it is intended to link the policy regime concept back to an academic debate in energy and mining policy, with a view to further substantiating a still somewhat 'elusive' term in that debate (Parsons and Moffat 2014).

To be sure, the pertinent SLO literature tends to focus on the relationship between corporations and local communities in the extractive industries, and the extent to which companies enjoy a social licence for their operations. This book, by contrast, investigates the creation of pertinent legal frameworks and their implementation. Still, there is an important conceptual link between the social licence and policy regimes for shale gas. A key notion in this context is 'legitimacy', a term featuring prominently in both scholarly debates. The SLO literature looks at legitimacy in terms of the non-legal aspects determining whether corporate activities are approved or accepted by social stakeholders. The policy literature discusses it in terms of stakeholder expectations on how policies should come about in the first place. In both cases, it is about participation, transparency and due process. Policy regimes can, as discussed, converge actor expectations; they may also enhance information flows and facilitate inclusive decision-making processes. With this, they may lend the very legitimacy to shale gas policy that creates the social licence it requires on various governance levels.

Clearly, the fracking technology has primarily local impacts; its social contestation, however, goes beyond community levels. In fact, debates surrounding shale gas tend to address fundamental – and, indeed, national-level – questions around whether shale gas is a resource that the country as a whole should embrace; whether the benefits of unconventional energy production justify potentially significant costs on other ends, such as risks to the environment and habitat; and the political economy behind accounting for the winners and losers of fracking, on all governance levels. What's more, most extractive techniques are embedded in existing legal frameworks; fracking, a novel technology by contrast, often requires at least amendments to national regulations and laws. It is therefore not only the local-level dynamics that determine whether fracking enjoys an SLO; the national level, in conjunction with the regional, also makes the difference. In other words, while the 'social licence to frack' clearly materializes on a local level, social acceptance needs to be (successfully) aggregated – and a social contract on fracking needs to be generated – at all levels of governance. This conceptual amendment is in line with several works stressing the importance of various constituent groups, networks of stakeholders and the wider society when it comes to creating an SLO (Boutilier and Zdziarski 2017; Gunningham, Kagan and Thornton 2004; Nelsen 2006).

The next three chapters delve into the case studies and explore shale gas policy regimes in Poland, Bulgaria and Romania.

4 *The Stalling Front Runner: Poland*

4.1 Polish Shale: Policy Context

Poland was for long considered the front runner in Europe's emerging 'shale gale' (Financial Times 2010; International Business Times 2010; The Economist 2011). This essentially is a function of the country sitting on 148 tcf or 4191 bcm of estimated reserves, Europe's largest (Figure 4.1) (EIA/ARI 2013), and of a persistently high support rate for unconventional energy among the population (Garpiel 2014; Polish Geological Survey 2014). The Polish government was determined to make shale happen, and eager to position the country as the go-to place in CEE for international investment in unconventional energy. Indeed, and even though the Polish Geological Institute (PGI) put domestic shale reserves at only 346–768 bcm (PGI-NRI 2012) – and hence much lower than the EIA estimates – Poland attracted great interest among foreign oil and gas companies. International majors such as ExxonMobil, Total, ENI and Chevron acquired licences, the prerequisite for shale gas prospection and exploration, as did mid-sized companies specializing in unconventional energy, including Talisman, Marathon, Cuadrilla and Lane Energy. Polish state-owned energy companies joined in and started joint projects with foreign partners. By 2015, investment in shale gas exploration totaled USD 2 billion (Financial Times 2015a). Poland, it was believed, would be the 'next frontier'.

And yet, at the time of writing, there is not a single well that produces commercial shale gas in Poland. To be sure, there is not one well in Europe that does, which reflects the fact that almost all the elements that facilitated the remarkably fast growth of the US shale gas (and later oil) industry are absent in Europe (see Chapter 2). Moreover, soft international gas markets and a fundamentally changing pricing environment soured the prospects for high-cost shale gas to compete against pipeline gas or even LNG. The oil price decline starting in 2014 effectively halved prices globally, which made many

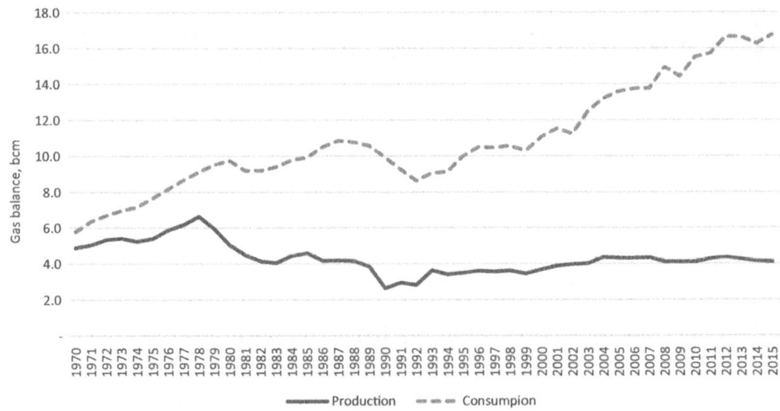

Figure 4.1 Poland's gas balance, bcm
Source: BP (2016)

companies shelve ambitious upstream investment plans and readjust their international growth strategies. European shale gas prospects suffered from this readjustment over-proportionally, and companies reacted by putting riskier and less promising projects on hold (Financial Times 2015a).

But Poland's shale gas sector also had to fight a tougher uphill battle than many had expected after the first shale gas exploratory licence was issued in 2007. Difficult geology dampened the high hopes put on Polish shale deposits. Reserves turned out to be deep and high in clay, which presented serious technical challenges for exploitation and diminished the prospects of producing unconventional gas at economic costs. Moreover, as will be discussed later, a poor regulatory environment and inadequate legal frameworks have widely been blamed for hindering shale gas development in the country. As a consequence, the annual number of exploratory wells drilled in Poland came down from twenty-four in 2012 to one in 2015 (Polish Geological Survey 2015a). Overall, fifty-three concessions were active by January 2015, and sixty-eight exploratory wells were drilled. The goal of 309 exploration wells by 2021, a target set by the Polish Ministry of the Environment (2015), is therefore very likely to be missed. In fact, all major international energy companies have given up their stakes in Polish unconventional gas, and most of their concessions have been picked up by smaller private companies and Poland's state-owned corporations

(Financial Times 2014c; Reuters 2013a, 2015b). As some observers argue (Zborowski 2015), it will now be on them to make Europe's 'first commercial shale gas' (Bloomberg 2014) become a reality. Clearly, the prospects of a sizeable Polish unconventional gas sector have become bleaker, and the enthusiasm characterizing the early days has given way to an acknowledgement of the harsh realities facing the country's shale industry going forward.

In terms of the policy context for Polish shale gas, history casts a long shadow. Poland's geographical location between two major European powers – Russia and Germany – clearly left a mark on national policy priorities until today. As a result of Poland's territorial integrity being historically threatened by those powers, retaining and strengthening national sovereignty is *raison d'état*, regardless of the political party in power. It is particularly Poland's high import dependence on Russian gas – around 60 per cent of consumption (Eurogas 2014) – that is considered a foreign policy threat (see Chapter 2). With this, and as explained later, the national security context also plays an important role in formulating energy policy goals. Poland is keen to diversify away from Russian gas and to tap new sources. Measures in this regard include interconnectors to the Czech and the German market in order to source molecules from the West, an LNG terminal in Świnoujście at the Baltic Sea coast and a planned additional pipeline project to source gas from Norway. Indigenous resources such as shale gas are part of the country's diversification strategy. In other words, foreign policy concerns play an important role in driving the country's shale gas policy.

Yet, it is not only foreign policy challenges that play an important role in Polish shale. In fact, against the backdrop of Poland's heavy reliance on domestically produced coal, EU climate policies are are susceptible to put pressure on the incumbent energy mix, as they force the country to replace coal with less pollutant energy sources – including renewables and natural gas. This is viewed by many as bearing energy security risks. Against an 88 per cent share of coal in Poland's power production (European Commission 2014f), EU decarbonization targets – and particularly the prospects of rebounding carbon prices – are perceived as significant economic risks for industry and households. Moreover, fears have emerged that EU climate policies could lead to additional imports of gas, and could thus aggravate the country's dependence on Russia even further. To be sure, Poland has made a name for itself as a key opponent to EU decarbonization policies, including any form of carbon pricing (such as through the EU Emissions Trading System, ETS), mandatory renewable targets or regulatory approaches to reducing emissions. Poland, in the past, dragged its feet when it came to EU carbon policies, and even vetoed the EU Low Carbon Roadmap 2050 (Bloomberg 2012; Euractiv 2012b, 2014). With EU decarbonization policies having come to entail binding emission targets, however, Polish energy policy will need to refocus on hedging the consequences for the economy and energy security.

Therefore, the country's primary energy policy goals are largely a matter of consensus across the aisle. This is remarkable given the deep political cleavages separating the two sides of the political spectrum.[1] One may even go so far as to state that while both parties' political goals seem to be at odds on virtually all fronts, energy policy is the one area where they tend to unite in their common support for fossil fuels and the importance of indigenous energy production, including shale gas. It is telling that notwithstanding the profound political differences between the centrist government of the Civic Platform (PO) (led by Donald Tusk, in office from 2007 to 2014) and the succeeding government of the nationalist Law and Justice Party (PiS) (under the leadership of strongman Jaroslaw Kaczynski), representatives of all parties also unite in maintaining Poland's opposition against any EU-imposed limits on shale gas extraction. As a corollary, Poland's pro-shale policy seems to be undisputed across the political spectrum. Poland's August 2014 Energy Security and Environmental Strategy defines shale gas (besides lignite) as key a priority for Poland's energy policy going forward (Ministerstwo Gospodarki/Ministerstwo Srodowiska 2014). Although the 2014 strategy was drafted under a PO government, the successor PiS government remains publicly committed to shale gas. This is notwithstanding the stronger role that the PiS administration assigns to Polish national energy companies in securing the country's energy supplies, the renewed focus on LNG imports and the creation of a new Ministry of Energy uniting energy and climate competences.

It is possibly because of the determination of the political elite that Poland remains the one country in Eastern Europe that it is still

[1] The Civic Platform (PO) effectively accused the Law and Justice Party (PiS) of a *coup d'état* because the new government had rolled back reforms and allegedly hollowed out the separation of constitutional powers (Reuters 2015c).

believed will eventually provide a roadmap for other CEE countries going forward, despite all the backslash the nascent industry has seen so far. Indeed, an Eastern European front runner will prove important for the future of unconventional gas in the region. CEE is markedly different compared to, say, the UK, where a sizeable market for gas was made subject to the liberal market regime and where there is a well-established regulatory framework in the extractive industry. Poland, by contrast, still remains vested in strong post-Communist institutional legacies in the energy sector. In fact, the country – and CEE more generally – remains a far cry from having well-regulated and liquid gas market structures akin to those in Western Europe. Heavy-handed regulation and incumbents preventing competition are typically named among the most important impediments to the development of more competitive markets for gas (EFET 2016). Yet, it is precisely these legacies and the still demanding market environment that could make the country a locus of regional experimentation for shale gas investment and governance, as they allow a 'contextualization' of the novel technology in a market environment that remains dominated by state-owned companies, significant red tape and limited administrative capacity.

The Polish shale gas story therefore offers important insights into the factors that enabled the government to push for and establish pro-shale gas policies in the country – regardless of their eventual outcome in the shape of a thriving or lagging Polish shale industry, or of production volumes. As the remainder of this chapter with explain, a strong policy narrative, combined with flexible institutional procedures and proactive outreach on the part of state authorities, ensured the buy-in of important stakeholders on national and subnational levels, even against the backdrop of limited administrative capacity and significant red tape.

4.2 Actors and Sector Governance

Actors and governance levels in Polish shale are numerous and multilayered. Principally, Poland's shale gas sector is subject to the same principles that inform regulatory frameworks in the energy sector more broadly. These are stipulated in the 1997 Energy Law, which defines the public bodies in charge of national energy governance and oversight, their responsibilities and the role of the country's national oil companies (NOCs); and in the Geological and Mining Act of 2011 (including its amendments of 2014), which sets the framework for issuing concessions and for conducting E&P activities. In addition, EU directives pertaining to competition policy and environmental protection, as discussed in Chapter 2, inform national energy policy-making. Following on EU pro-market policies, the Polish gas industry was gradually liberalized and deregulated. Still, the Polish state retains a strong role in the energy sector. National energy companies and semi-state-owned corporations such as LOTOS Group, PKN Orlen and PGNiG – some of which constitute vertically integrated corporations that cover the entire energy value chain – remain dominant players in the domestic market. In upstream gas, incumbent PGNiG, the Polish Petroleum and Gas Mining Company, dominates the market with a 98 per cent share (IEA 2014a). PGNiG traditionally also handles all imports from Russia through the Yamal Europe pipeline, which in 2015 amounted to 9 bcm (Gazprom Export 2016). This is likely to change, as the Polish government is on record stating that it will not renew its long-term supply contract with Russia, which dates back to 1996 and ends in 2022 (Reuters 2016b). In addition, the Świnoujście LNG terminal will bring up to 5 bcm of gas into Poland going forward, with a 1.5 bcm long-term LNG contract with Qatargas in place (which gas will also be marketed by PGNiG). Polish NOCs, finally, represent some of the country's largest corporations. To put the Polish energy industry (including electricity and heat) into perspective, the sector's revenues amount to some 20 per cent of the country's GDP (EMIS 2014). This makes it an important stakeholder not only in a potentially emerging shale gas industry, but also in the country as a whole. International oil companies (IOCs) play a role mainly in Polish gas trading, although PGNiG also retains a dominant share here. In addition, transmission operators PERN (oil) and Gaz-System (gas), and separate pipeline operators such as EuroPol-Gaz, operate grids and infrastructure.

On the federal level, several public entities are charged with energy sector oversight.[2] The Treasury is the legal owner of hydrocarbon resources and oversees Poland's NOCs, as well as the government's stakes in PGNiG (72 per cent) and other semi-state energy companies.

[2] This paragraph refers to the period up to 2015, i.e. the time period of this investigation. The creation of the Ministry of Energy in 2015 brought about an overhaul of the country's governance structure.

The Ministry of Environment, by contrast, grants licences and concessions, and resumes environmental oversight. The Ministry of the Economy, although less central to the shale gas sector, is in charge of drafting the country's energy policy, as defined by the 2014 Energy Security and Environmental Strategy (Ministry of Economy and Energy 2014). In addition, the Ministry of Foreign Affairs retains an important role in coordinating international outreach and the external dimension of Poland's efforts to develop a domestic unconventional gas sector. This notably includes the exchange with foreign representatives from politics and business, particularly from the US. A major venue in this regard was the previously mentioned 'Global Shale Gas Initiative' of the US State Department, now termed the 'Unconventional Gas Technical Engagement Program', which since 2010 has sought to foster technology transfer and promote hydraulic fracturing technology, particularly in Eastern Europe. Several regulatory agencies perform functions related to sector management and monitoring, including the approval of natural gas tariffs (Energy Regulatory Office, URE) and anti-trust issues (Poland's Office for Competition & Consumer Protection, UOKiK).

In addition to the federal level, regional (*voivode*)-level agencies exert authority in fossil fuel production and hence shale gas matters. Sixteen Regional Directorates for Environmental Protection (RDOS) – the *voivode*-level arm of the Ministry of Environment – implement and monitor provisions related to environmental protection, including the environmental impact assessments required in mining and other extractive activities. These are complemented by the General Directorate for Environmental Protection (GDOS), which reviews RDOS activities and acts as the body of appeal. Moreover, local authorities perform important roles in energy upstream by overseeing (and granting) water resource rights and waste management, and by performing consultation during the licensing process. As a consequence, environmental procedures and approval processes relevant for extraction activities crucially involve the regional and local levels. Fiscal regimes reflect also the strong role of subnational entities: the mineral taxation regime allocates only 40 per cent of revenues to the national level, with the majority – 60 per cent – going to local levels. As we shall discuss later, this complex and multilevelled governance framework presents opportunities, but also challenges, for shale gas policy.

Relevant socio-economic actors include environmental NGOs. Although civil society organizations in Poland are not as deeply embedded socially or as organized as they are in Western Europe (Fagan and Carmin 2011), environmental groups have formed across the country. The oldest group, the Polish Ecological Club (PKE), has its roots in the Solidarity movement and operates fourteen regional offices and 120 'grassroots circles'. Other important organizations include the Polish Climate Coalition, which represents twenty-three environmental NGOs, which typically are very localized in nature. It is important to note that Polish NGOs are not explicitly formed around shale gas, but cover issues as diverse as Polish coal, emissions and pollution stemming from heavy industry, and climate change policies. Because these groups are embedded in local contexts, they have significant mobilization capacity – as became obvious in the case of Zurawlow, a village where, as we shall discuss later, activist groups successfully mobilized against planned shale gas exploration. Additional important economic stakeholders are represented by business associations such as PKPP Lewiatan, the Association of the Polish Oil & Gas Upstream Industry, the Polish Employers Association, industry chambers and trade unions.

In sum, shale gas policy in Poland needs to account for multiple governance levels and a diverse set of institutions and actors characterizing the country's energy sector. The next section assesses the institutional arrangement pertaining to Poland's shale gas story.

4.3 Poland's Policy Approach: Institutionally Path-Dependent, Inclusive and Unwillingly Experimentalist

Poland's institutional approach to shale gas can hardly be called one of great design. In fact, three elements characterize the organizational arrangement pertaining to shale gas policy: strong path dependence, with all the institutional baggage that comes with it; inclusive processes regarding key stakeholders; and – albeit unwillingly – a good dose of regulatory experimentalism, which may not please investors, but which facilitates learning by doing on the ground.

The way Poland chose to govern shale was by building on the existing institutional structures in the extractive industries. In essence, Poland used the administrative apparatus regulating conventional hydrocarbon extraction in coal, oil and gas, an apparatus which itself had to work within the remits of pre-existing legal frameworks.

Clearly, this came with an institutional legacy. The country's relevant legal frameworks had been established for an oil and gas industry dominated by state-owned companies and an energy sector organized around the public utility model. As a consequence, they fell short of the practices characterizing more liberal models of energy sector governance. This point was vividly illustrated by Poland granting hydrocarbon concessions to select companies without carrying out the compulsory public tender procedure, and failing to guarantee non-discriminatory access in the realm of prospection, exploration and extraction of hydrocarbons, thereby openly violating EU law (European Court of Justice 2013). More importantly, perhaps, Poland's institutions of energy governance were not designed to pay attention to the profit motive or the regulatory impact on the corporate cost structures which determine investment decisions by foreign companies. Yet, these foreign investors are crucial to developing Polish shale, due to the capital they bring and their technological know-how. While domestic Polish corporations have considerable experience in prospecting and producing conventional gas (and oil), they lack the expertise to master fracking, a complex technology. In fact, as various representatives of the energy service industry suggested during background interviews conducted for this book, fracking resembles a manufacturing process rather than a 'typical' drilling job in conventional gas. Fracking therefore requires different managerial and engineering skill sets than conventional gas – the sector Polish corporations traditionally focused on.

What's more, energy policy-making in Poland prominently features what the public policy literature has termed policy 'stovepipes' or 'silos' – clearly demarcated areas of (regulatory) activity with little outreach to neighbouring administrative units or policy fields – which has prompted calls for 'joined-up government' as an integrated policy strategy (Rayner and Howlett 2009). Yet, as discussed in Chapter 2, shale gas cuts cross clear lines of authority, as typically pre-established in national systems of energy governance – including in Poland. Institutional initiatives to improve the stovepipe situation, notably through the 'Interministerial Team for implementation of Energy Policy of Poland until 2030', created in 2010 with a view to coordinating national-level activities in the sector, have not delivered (Godzimirski and Kasianiuk 2014). In light of this, some of the long-established institutional patterns – both in terms of regulatory frameworks and with regard to the organizational set-up of supervising authorities – do not meet the specificities of shale gas as a sector. More to the point, they are at odds with the needs and expectations of the foreign players that bring much-desired technical expertise, specialized knowledge and capital.

Institutional legacy has given rise to several sets of problems in Polish unconventional gas governance. The first relates to policy coordination and administrative capacity. A function of pre-existing silos in Polish energy policy, administrative coordination across regulatory authorities was lacking in Polish shale (interview with Political Advisor, Instytut Studiow Energetycznych, 2012). For instance, as the Polish Supreme Audit Office objects in a 2014 evaluation, no governmental body was put in charge of coordinating and monitoring relevant economic and legal aspects of shale gas exploration across relevant units (Supreme Audit Office of Poland 2014). This clearly presents a significant challenge for private companies that have to operate in an environment characterized by administrative structures that are not capable of meting their expectations regarding licensing processes or the issuing of necessary permits. As put by a private industry representative, global players are key for Polish shale, and if they are not convinced within '5 minutes, they will go somewhere else' (interview with Country Manager of United Oilfield Services, a Poland-based oil and gas services firm, 2012). A manager of an international energy company specializing in shale gas echoed the prevailing 'bureaucratic lag' and suggested that relevant authorities remained seriously understaffed to deal with at times more than 120 pending applications for concessions and licences (interview, 2012). This ties into industry complaints over 'excessive bureaucracy' characterizing shale gas exploration permits, and administrative procedures more generally (Natural Gas Europe 2013a). Moreover, as energy industry representatives stressed, the development of the Polish shale sector would need to be brought in sync with broader developments of the energy sector as a whole (interview with Chief Economist, Strategy and PPM Department, PKN Orlen, 2012). Yet, while the government pushed shale gas as a policy priority, the institutional structure did not see the necessary updating, nor did top-level state authorities reveal an interest in bringing administrative processes 'up to speed'.

To be sure, upstream regulation itself was often criticized for being restrictive, particularly with a view to the requirements pertaining to environmental impact assessments (an issue we shall return to later). For instance, the Chief Economist of the Polish state-owned energy company PKN Orlen mourned that '[w]e have a more restrictive regulatory environment than the EU or US. [The authorities] ask how deep do you go, and then if you want to go 100 meters deeper, you need to apply again' (interview, 2012). And yet, regulation itself does arguably not present the biggest problem as perceived by industry representatives. In the words of a manager of an international energy company specializing in shale, 'environmental regulations are extremely strict [but] we are fine with this as long as they are clear' (interview, 2012). By contrast, other interviews revealed that the more fundamental challenge lies with the lack of regulatory coherence and streamlined procedures, which emerged as an issue for all stakeholders in shale gas, 'investors and social organizations, including commercial organizations' (interview with Regional President of Ecological Club, 2013). As even representatives of environmental NGOs admit, '[w]hat is difficult is every environmental assessment is different. They can be different in information and quality', an issue which 'has to be solved by improving the quality of administration' (interview with representative of Climate Coalition, 2013). Seconding this, corporate representatives stressed the need to enhance administrative procedures and clarity in environmental permitting processes, rather than necessarily relaxing environmental provisions.

Adding to the red tape and lack of institutional coordination, relevant state authorities were in fact hindering rather than advancing shale. As Poland's Supreme Audit Office concluded in an internal evaluation, 'indolent' public authorities obstructed private entrepreneurship in unconventional gas, thereby aggravating a deteriorating international market environment (Supreme Audit Office of Poland 2014). The report found that shale gas regulation was caught in questions of inter-ministerial authority, that there was no clear point of ownership for unconventional hydrocarbon development, that key ministries (notably the Ministry of Environment) did not prioritize shale gas, that personnel was insufficient to deal with the licensing processes and that investors were not treated equally. In short, all essential elements characterizing a joined-up approach were absent. This, as the audit office found, slowed the exploration process in the country 'at least for several years' (Supreme Audit Office of Poland 2014). While these findings clearly suggest a severe lack of institutional capacity, they also reflect the fact that shale gas had been made subject to pre-existing governance structures, administrative processes and regulatory frameworks that were ill-designed to properly handle the novel technology and its specificities.

A final element pertaining to institutional legacy relates to the various frictions surfacing during regulatory adjustment procedures. The government had taken several initiatives to amend pertinent laws and regulatory frameworks. Yet, persistently scattered institutional competences across relevant ministries caused delay, as efforts to change legal frameworks effectively 'stopped at the stage of inter-ministerial agreements' (Supreme Audit Office of Poland 2014). This resulted in regulatory inconsistency. Examples include the long-pending fiscal framework for shale, which in 2014 materialized in the shape of the Act on a Special Hydrocarbon Tax. Although it stipulated tax breaks for several years, the Act was met by criticism on the part of the energy industry and among commentators (Polish Geological Survey 2015b; Wall Street Journal 2014). It was viewed as premature and as setting the wrong incentives for a nascent industry. The 2014 Act was eventually emended by the 2015 Hydrocarbon Investment Act. Another case in point is Poland's 2013 attempt to streamline environmental impact assessment regulations in shale gas. The regulatory overhaul was carried out with the aim of clarifying the status of shale gas investments in assessment procedures. As Poland had channelled shale gas into the existing legal frameworks, this warranted altering regulations for conventional oil and gas exploration. A key issue in this regard was exploration above a depth of 5000 m (where much of the shale gas is located) and in areas not subject to special environmental protection (e.g. Natura 2000 areas). Here, a 'simplified' assessment would replace the 'comprehensive' one that had applied to all drilling activities previously. For environmentally sensitive areas, or areas in close proximity to residences, the RDOS can still request a comprehensive assessment. Production activities, by contrast, always prompt a full assessment (interviews with Director, Regional Environmental Agency, 2012; Energy and Natural Resources lawyer, White & Case Poland, 2015).

As observers noted, the alterations in environmental impact assessment regulations were in part a reaction to the uncertain environment companies had to operate in when prospecting for shale gas (interview with Energy and Natural Resources lawyer, White & Case Poland, 2015). The amendments, however, clashed with EU regulations, which prompted the European Commission to start an infringement procedure against Poland in 2014, on the basis of Poland contravening the Directive on Environmental Impact Assessments (Neslen 2014). To be sure, EU queries into the legality of national regulatory provisions are nothing unusual. The important point here is that the legacy in the Polish energy governance system in the shape of agency stovepipes prevented a comprehensive revision of the country's legal frameworks so that it would do justice to energy sector needs and simultaneously satisfy EU requirements.

While the institutional set-up indeed presents various challenges to the design and coherence of Polish shale gas policy, the policy approach proved inclusive towards key stakeholders. For one thing, the Polish government ensured that national energy companies became engaged in the unfolding 'shale gale' and had a role to play therein. To this end, the Polish treasury – the owner of the country's NOCs – obliged national key players such as PGNiG in natural gas, PGE Polska Grupa Energetyczna, Enea and Tauron Polska Energia in the utility sector and KGHM Polska Miedz in the mining industry to form joint ventures with foreign energy companies (Wall Street Journal 2012). To be sure, a key motivation for this step may simply have been to avoid a repetition of the experience related to big-bang privatization programmes in the 1990s, which in many post-Communist societies are still regarded as a sell-out of national assets. Arguably, however, this approach also tied Polish state-owned corporations firmly to shale gas projects in the country, made them stakeholders and gave them an interest in making shale a success. Further, it opened up opportunity for technology transfer in unconventional gas, which benefited PGNiG and its like, and gave them a quick learning curve in a sector that was new to them.

This approach also made sure state-owned companies shared in the potential revenues from shale gas extraction. As non-gas companies were brought in to the process, this entailed a certain element of competition for the incumbent position of PGNiG. As representatives of foreign companies mourned, the role given to Polish state-owned companies may even have been too dominant, which sometimes hindered rather than helped exploration activities, not the least due to the Polish NOCs coming with the mindest of public monopolies, not of private corporations (Reuters 2013c). Yet, co-opting Polish NOCs facilitated the buy-in of this important group of incumbents in the Polish energy sector. In other words, the power arrangement was comprehensive and included corporate actors who may have had an incentive to act as veto players the moment shale gas was about to start threatening market share or revenue streams.

For some time, Warsaw had toyed with the idea of setting up a National Operator of Energy Minerals (Narodowy Operator Kopalin Energetycznych, NOKE), a plan which had sparked controversy among observers. As some have argued, NOKE would have added regulatory and managerial coherence to Poland's energy sector, and remedied institutional fragmentation (Godzimirski and Kasianiuk 2014, 26). Although the plan to create a Polish oil and gas NOC was not pursued further by the PO government, the topic re-emerged in the wake of PiS's gaining power in 2014, and the new government openly pondered merging PKN Orlen, Lotos and PGNiG (Reuters 2016b). Yet, at the time, the decision to refrain from establishing NOKE and rely on existing NOCs clearly served the purpose of fostering institutional 'outreach' and came with the benefit of providing stakeholder inclusion.

Turning to subnational-level dynamics, an important process of community engagement unfolded on a local level. To be sure, this process was not based on a nationwide 'master plan', and relied on the initiative of regional authorities. Importantly, however, it was sanctioned by the national-level administration. A key element in this context was information. As stressed by an advisor to the Minister of the Environment, 'the most important factor is to provide the information to people so they can take decisions based on information from coming at least from two sources if not three' (interview, 2012). Here, information was seen as a two-way process. Companies were encouraged to hold town hall meetings or other types of public fora in order to facilitate exchange over concerns the local population might have and to 'clarify the situation, so it encourages people to participate' (interview with Deputy Director, RDOS, 2013). Legally, companies were not obliged to do more than release public notices of drilling applications. Yet, as interviews revealed, they were expected to reach out to and engage with local communities, and hence to go beyond what was formally stipulated in legislation or regulation. In fact, for the Polish Treasury, it was 'very important for companies to engage with the local population [and] to have local dialogue' (interview with for a Polish Treasury Official, 2012). In its own words, the motivation for the Treasury – the owner of state oil and gas companies – was to remain 'invested in how companies engage with local people' (interview with Polish Treasury Official, 2012).

Besides the stated intent of regional- and national-level authorities to tie local communities into the policy process, there clearly also exists an institutional imperative to do so: local authorities play an important role in approving exploration permits and in facilitating the surrounding infrastructure. This gives them leverage. More to the point, companies have to work and coordinate with RDOS, GDOS and local authorities during the application process for shale gas exploration. In fact, it is local authorities that first develop a plan for exploration activities. RDOS – the authority overseeing environmental impact assessments – then consults with relevant local authorities, before issuing a decision on environmental aspects. Finally, local authorities are again asked for their opinion before the concession is granted to an investor or company (interviews with Energy and Natural Resources lawyer, White & Case Poland, 2015; Advisory to the Polish Ministry of the Environment, 2012; Deputy Director, Regional Directorate for Environmental Protection, 2013; and representative of Climate Coalition, 2013). As part of the process, investors are asked to share an information chart to both the local municipality and the regional environmental directorate, in order to detail their planned exploration activity (interview with Deputy Director, Regional Directorate for Environmental Protection, 2013).

As already indicated, local communities are also formally part of the revenue stream: regional and local authorities receive 60 per cent of mining royalties – which typically are a large chunk of overall hydrocarbon revenues – while the remaining 40 per cent goes to a National Fund for Environmental Protection and Water Management. In a move to overhaul tax regulations pertaining to shale gas, the government, in 2014, adopted a legal package that amended the Geological and Mining Act and introduced a new 'Act on Special Hydrocarbon Tax' (KPMG 2014). While the special tax will exclusively benefit the State Treasury, the amendments to the Mining Act impact on local income levels such that overall royalty levels will rise. This was meant to improve the profit local communities make from the potential revenue stream stemming from unconventional gas – in addition to revenues stemming from lease agreements, permit fees and other areas.

To be sure, efforts to build an inclusive process, maintain local outreach and ensure information sharing with the population did not necessarily happen everywhere in the country, nor were they made consistently and across the board. Overall, however, municipalities and villages enjoyed a relatively well-established 'institutionalized voice' in the shale gas concessions procedure. The institutional and regulatory framework also stipulated the benefits local communities could expect from shale gas extraction, if the prospection yielded promising results, and ensured that local communities would receive due information on planned drilling activities and the potential risks that these activities entailed. The information itself may not be unambiguous, as it is provided by companies with a material interest, and the sheer availability of data may be insufficient if it is not properly 'decoded' and made accessible for a lay audience. For this reason, several interviewees stressed the importance of public fora, involving the local community, experts and the investor (interview with Deputy Director, Regional Directorate for Environmental Protection, 2013).

All in all, it is fair to conclude that Poland failed to adopt a joined-up approach to shale gas policy. The continuous back-and-forth process that tied national regional and local authorities into the concessions procedure was complex and cumbersome, and it remained ad hoc to the extent that informal procedures often complemented (poorly designed) institutionalized ones. In that respect, Poland's shale gas governance may to a certain degree qualify as 'unwillingly experimentalist' (see Sabel and Zeitlin 2012). Unsurprisingly, therefore, the Polish institutional set-up did not generate great enthusiasm among observers (The Economist 2013b). Indeed, as the preceding analysis shows, it was lamented that more went wrong than right. From an analytical point of view, however, the Polish power arrangement co-opted important local and national stakeholders into the government's shale gas policy. It did so by way of ensuring that the procedural arrangement fostered information flows, institutional outreach and community empowerment.

4.4 Policy Narratives: Security, Economy and Jobs

Poland has seen a lively policy debate on shale gas. Compared with other European countries, its discussion centres much less on ecological concerns. Instead, two narratives have emerged as dominant in public debates. The first one, the *national security narrative*, frames shale gas in the context of Polish history and national integrity. This narrative is epitomized by a statement of then-Prime Minister Donal Tusk, who posed that 'gas security is a fundamental prerequisite of sovereignty' (Wall Street Journal 2014). In this context, shale gas emerges as a means to ease Moscow's perceived energy stranglehold, ensure reliable energy supply and reduce Poland's heavy reliance on Russian gas imports. According to Tusk, '[h]ard coal and lignite – and soon shale gas – will remain our principal energy sources. That's where the future of the [Polish] energy sector lies' (AFP 2013). With this, the national security narrative embeds Polish shale gas firmly in the context of geopolitics.

The national security narrative resonated broadly with stakeholders from the business community and the state apparatus, on all governance levels. As a representative of the Office of the Minister of the Treasury stressed, 'shale gas [is] part of [Poland's] diversification policy' (interview, 2012), a statement which was echoed by members of scientific community to the effect that shale gas production 'would be a milestone to be independent from Russia' (interview with professor of Polish Institute of Soil Science, 2013). On a ministerial level, it is the prospect of 'stable supply of gas in Poland [without] any political strings attached' that motivated pro-shale gas policies (interview with member of Economic Policy Department of the Polish Ministry of Foreign Affairs, 2012). Similar statements from representatives of the business community echo that view. As explained by a manager at United Oilfield Services, a Polish service company, 'energy is a foreign policy tool for Russia [and] shale gas opens up the possibility of being more secure from Russia's monopolistic position' (interview, 2012). Moreover, the manager pointed to the 'painful history for Poland under Russia' (interview, 2012), a statement which refers to Poland's trauma of 1939, when the country was partitioned by Soviet Russia and Nazi Germany, and later made part of the Warsaw Treaty bloc.

In addition to the energy industry and the state apparatus, society as a whole also bought into the security narrative. This is reflected by various domestic polls suggesting that the majority of the Polish population endorses shale for its perceived security benefits (see, for instance, the surveys discussed in Lis et al. 2015). Turning to environmental actors, neither the Green Party nor Polish NGOs would necessarily support shale gas production as a way to reduce Russia's share in gas imports and enhance the country's energy security. In fact, activists suspected the business community of utilizing the security narrative for economic ends rather than for political ones. A representative of

Climate Coalition shared that '[t]he way government talks about it as the perfect solution with dependence on Russia . . . We don't see it this way, and not even the international companies see it this way' (interview, 2013). But even among green activists and Polish environmental NGOs, the aspect of 'sovereignty is important' (interview with representative of Cleantech Poland, a consultancy firm active in the unconventional energy sector, 2012). As the Chairman of Poland's Green Party admits, '[c]itizens see the opportunity to be independent . . . from Russia' (interview, 2013).

The security narrative is complemented by a strong second frame, which pertains to *economic opportunity*. This frame embraces shale gas as a source of economic welfare and material benefit for the country. It is best represented by a statement of Radoslav Sikorski, Poland's former Foreign Minister, who alleged that shale gas could make Poland 'a second Norway' (Kenarov 2012). Key elements of this policy narrative include the perceived competitive edge for the manufacturing industry due to low energy prices, the effects on job creation and additional state revenues on national and subnational levels.

Like the national security narrative, the economic opportunity frame has resonated among constituencies across the society and economy. Naturally, the extractive industry, in the shape of the Association of the Polish Oil & Gas Upstream Industry, publicly supports the economic opportunity frame (Lis et al. 2015, 20). For industry representatives, shale gas 'can be important drivers for the Polish economy [because it] can produce cheaper gas for other economic sectors [such as] the chemical industry' (interview with director-level representative of the Polish Confederation of Private Employers, PKPP Lewiatan, 2012). Lower energy costs clearly also form part of the motivation driving the state apparatus, in addition to a 'knock-on effect on the energy consuming industry' (interview with former advisor to the Polish Foreign Minister, 2012). For the extractive industry, shale gas promises new fields of activity and additional revenue. As the chief economist at PKN Orlen explained, '[w]e go in because we believe in it and put money on our bet that we will find gas' (interview, 2012). Indeed, a study commissioned by PKN Orlen sets investment into shale gas production at up to USD 11.1 billion until 2025 (Czyzewski, Bodnari and Kozieja 2012) – in its moderate scenario. Much of the consequent investment needs would arguably be met by foreign capital flowing into the country. Hinting at the company's state ownership structure, PKN Orlen's chief economist insists that '[i]t is not the case that the government pushes us into this' (interview, 2012). Instead, it is the prospects of foreign direct investment, technology transfer and new skill sets for the company that motivate incumbent energy companies to get active in shale gas. Here, the government's decision to make state-owned corporations part of private industry-led gas exploration arguably exerted impact on corporate strategy.

In turn, the potential size of the Polish shale reserves makes major international oil and gas companies look at Poland 'as a base for their European oil and gas business' (interview with former advisor to the Polish Foreign Minister, 2012). The Treasury, overseeing state-owned companies, therefore aims 'to encourage investors to bring money here and drill and drill ... We will do our best to help investors, because our companies will benefit as well' (interview with representative of the Office of the Finance Minister, 2012). As confirmed by industry representatives, the perceived economic impact of an emerging shale gas industry motivated the state administration and political leaders to do 'everything ... to start a program quickly' (interview with director-level representative of PKPP Lewiatan, 2012). This even applies to the ministry in charge of environmental oversight. According to an Advisor to the Minister of the Environment, environmental authorities are aware of the financial risks facing investors, 'so we will try to regulate it in a way that will boost the level of investments and security of investments' (interview, 2012).

For government representatives, it is particularly local communities that would stand to benefit from unconventional energy production. In the words of a member of the Economic Policy Department of the Polish Ministry of Foreign Affairs, 'shale gas [is] a tremendous opportunity to speed up their development' (interview, 2012). This perception is confirmed by polls generally suggesting that the majority of Poles support shale gas because of its economic benefits, as well as by surveys among local authorities, who overwhelmingly expect additional revenues and a boost for the local development of small and medium enterprises (Lis et al. 2015, 11). Interestingly, although polls suggest a NIMBY effect at work, Polish citizens express support for shale gas production in their area of residence by 66 per cent (Polish Geological Survey 2014). According to a 2013 EU report, Polish public opinion even favours 'unconventional fossil fuels such as shale gas as an energy option' over conventional fossil fuels such as coal or natural gas, by

32 per cent. This contrasts with other European countries, where, on average, 9 per cent of the population thinks shale should be a priority – in some countries, it is only 3 per cent (European Commission 2013c). These data suggest that the economic opportunity frame resonates down to local levels.

Importantly, this narrative extends to what amounts to an insurance policy against EU decarbonization goals. To be sure, and as discussed earlier, Poland remains vocal in its opposition to EU climate policies, and the country keeps dragging its feet on European decarbonization targets. And yet, EU climate polices are understood as remaining a persistent medium- to long-term threat to the economy and the Polish industry, in the shape of significantly higher carbon prices and hence rising electricity bills for households and businesses (interview with representatives of the Polish Treasury, employers' association and oil and gas companies, 2012). Shale gas, therefore, not only presents a hedge with regard to Polish energy prices, which are mourned by business representatives as being among the highest in Europe (interviews with manager at United Oilfield Services, 2012),[3] but may allow the carbon footprint of the Polish energy mix to be lowered, shielding the economy from potentially negative economic effects stemming from the ETS and EU carbon policies.[4] It is in this context that the Polish science community promotes shale gas (and natural gas more generally) as a 'bridging fuel for renewables' (interview with scientist at the Polish Institute for Sustainable Development and a member of a county-level climate project, 2013). Prominent scientific institutes, including the PGI, have taken on the task of flanking Poland's emerging shale debate with detailed scientific expertise (and produced their own estimates of the country's reserves). The view of shale gas supporting energy transition by way of supporting a soft landing in a low-carbon future is broadly shared by the business community. Shale gas, as an industry representative argued, 'would be a good source of

[3] Although disputable with a view to retail and industrial gas and electricity prices, this claim is often repeated in the context of the Polish energy debate (see, for instance, PKEE 2016).

[4] It is important to note that the carbon footprint of shale gas along the value chain is not necessarily better than that of coal. Various studies point to a problematic record of shale gas in this regard (Alvarez et al. 2012; Brandt et al. 2014; Howarth, Santoro and Ingraffea 2011). The statements made here pertain to the views as expressed among Polish interviewees.

energy, [and] it would support renewables' (interview with director-level representative at PKPP Lewiatan, 2012).

Still, the technology remains contested among environmental groups and parts of the local population, despite the generally positive narratives surrounding shale. Scepticism relates to environmental concerns, groundwater safety issues and the potential risks to local habitat stemming from unknown ingredients of the fracking fluid. Environmental NGOs have publicly raised the alleged negative side effects pertaining to hydraulic fracturing, and have started to organize locally. The prospects of gas exploration using vertical hydraulic fracturing led to protests in various communities across the country, with the village of Zurawlow representing the most prominent and publicized case. Here, local protesters, with the help of activist groups, mobilized against planned drilling activities by Chevron, the US energy major, and made the company abandon its activities in the region (The Guardian 2015). Protest also formed in the Pomorze region, in Baltic seaside areas where shale gas concessions had been granted and in the Kashubian Lake District (Zielinski 2012). Some of these protests had reported links to international anti-shale movements and organizations, including Food & Water Watch Europe, No Fracking France and UK movements, as well as to German and EU-level MPs (Lis et al. 2015, 22). Still, few rose to similar prominence or escalated to the same levels as Zurawlow. Protests also never reached the national level, and an 'environmental bane' narrative capable of resonating among broader parts of the population did not emerge in Poland, in stark contrast to Bulgaria (see Chapter 5) and other EU countries. In this context, it is telling that PKE, Poland's most prominent national-level environmental organization, abstained from engaging in anti-fracking activities. This may be due partially to the Polish government trying to legally curb NGOs from getting involved in shale gas matters (Natural Gas Europe 2013b). More importantly, it is because adopting a critical position towards shale, for instance based on environmental concerns, became difficult in light of a strong security narrative and because 'any objection is stigmatized as pro-Russian' (Zielinski 2012).

Moreover, even the environmental community cautiously started to embrace unconventional gas as a means of decarbonizing the Polish energy system. It did so reluctantly, but motivated by the fact that '[w]e believe [a] local use of shale gas [utilizing the] best available technologies could be a transition fuel that could complement the use of renewables' (interview with representative of Climate Coalition, a Polish environmental NGO, 2013). Therefore, '[s]hale gas is not a priority for our organization, [as] we look at the whole energy mix' (interview, 2013). Unlike in most of Western Europe, environmental groups did not necessarily adopt a position on shale gas – the Environmental Protection League Lublin is a case in point (interview with the President of the Board, 2013). Further, while it opposed shale, the Polish Green Party made efforts to point to alternative options, including energy efficiency measures (interview with Chairman of Poland's Green Party, 2013). Finally, members of the energy industry certify that Polish environmental NGOs primarily stressed that extraction needed to be carefully monitored, rather than precluding the option of shale gas exploration altogether (interview with corporate affairs manager, Talisman Energy, 2012). In sum, '[w]ithin environmental organizations, there is a pragmatic understanding' (interview with representative of shale gas consultancy firm Cleantech Poland, 2012). This, indeed, clearly distinguishes the Polish situation from those in other European countries, and it suggests the 'economic opportunity' narrative received support from civil society.

4.5 Assessing Poland's Policy Regime: Comprehensive Power Arrangement, Low Institutional Capacity and High-Valence Narratives

Overall, the Polish energy sector is characterized by multiple governance levels and a diverse set of actors. These characteristics, by extension, also pertain to the nascent shale gas sector. Key stakeholders exist among ministries, industry, businesses, societal actors (including NGOs) and local communities. Not all of these stakeholders may be able to exert formal veto power in the policy process, as they are not given an equally institutionalized 'voice' in that process. Yet, due to either their sheer size and importance to the economy (e.g. NOCs) or the degree to which they are able to mobilize resistance and gain media coverage (e.g. local communities), even non-shale industry players and municipalities emerge as key actors in the shale gas policy process. The Polish power arrangement accounted for this diverse set of interests, and notably co-opted national energy companies into the process, turning them from potential veto players into stakeholders.

With regard to the procedural arrangement, Poland's approach to shale gas policy clearly did not live up to the ideal of 'joined-up

Table 4.1 *Summative assessment of Polish policy regime in shale gas*

	Analytical focus	Key indicators (high/low)	Value
Ideas	Valence of policy frames	• Degree of 'ideational uptake' among stakeholders	• High
		• Congruence of dominant interpretative frames of governmental actors and societal groups	• High
Interests	Constituency representation	• Involvement of stakeholders in the policy process	• High
		• Inclusion of veto players in the policy process	• High
Institutions	Communication	• Procedural outreach towards non-state actors	• High
		• Transparency of the policy process	• Medium
		• Information flow among stakeholders	• High
	Joined-up government	• Cooperation of relevant state authorities on all policy levels	• Low
		• Policy ownership across policy levels	• High
		• Flexibility in adjusting procedures and processes	• Medium
		• Administrative quality	• Low

government'. Instead, limited administrative capacity was coupled with inherited institutions designed for a conventional, state-run energy sector, significant red tape and a lack of coordination between relevant state authorities. At the same time, institutional procedures and proactive outreach on the part of state authorities ensured the buy-in of important stakeholders on the subnational level. This included experimenting with various formats of community engagement on the local level, notably with a view to enhancing information flows. This approach, often informal and even ad hoc, flanked the (often insufficient) formalized administrative procedures and fostered the involvement of local communities, municipality-level authorities and other stakeholders in the implementation of the government's stated policy goals. To a certain degree, it therefore made up for the significant institutional legacies that came with a sectoral approach to shale gas governance.

In terms of policy frames, the 'national security' frame and the 'economic opportunity' narrative, as promoted by the government, clearly resonated strongly among key stakeholders. The narratives were of high valence, and the data revealed a high degree of uptake among societal and business actors. This uptake extended even to the environmental community and NGOs – a rather unique situation compared to other European countries, or even the US. Moreover, strong policy narratives lifted shale gas to what amounted to a 'national project' in terms of enhancing the country's sovereignty and economic prospects. It therefore became hard for opponents to voice concerns or publicly resist policies fostering a domestically available unconventional resource base.

In all, as summarized in Table 4.1, Poland's shale gas policy regime is comparably strong. It offers 'buy-in' opportunities for key stakeholders, unites them around an overarching policy narrative (security gains and economic development) and thus lends legitimacy to the policy goals as stated by Poland's government.

5 *The Naysayer: Bulgaria*

5.1 Bulgarian Shale: Policy Context

The case of Bulgaria epitomizes some of the most pressing energy policy challenges in Eastern Europe, and the region's energy woes in natural gas. First, the country is completely dependent on Gazprom as the single source of gas supplies, and its structural dependence on Russian gas imports firmly ties the country's energy sector to its dominant external supplier. The January 2009 gas cut-offs, following on the Russia–Ukraine dispute over trade and prices, painfully revealed Bulgaria's significant exposure to external supply risks. In the midst of a harsh winter, the country was proverbially left out in the cold and came to a dead stop – a function of a lack of alternative supply options (Kovacevic 2009; Stern, Pirani and Yafimava 2009). The economic impact caused by the two-week Russia–Ukraine gas standoff amounted to a severe shock, and has been estimated at a loss of some 23 per cent of GDP for the period of the supply interruption, or roughly 1 per cent of GDP for the year 2009 (Christie 2009). Bulgarian sources cite BGN 500, or roughly EU 250 million, as the overall price tag (Energy and Water Regulatory Commission 2010). Energy security, therefore, features prominently in Bulgaria's national security strategy (State Agency for National Security 2011), and the Bulgarian government has stated its strong intention to significantly reduce Russian imports (ICIS 2011). Still, and despite EU policies fostering gas pipeline infrastructure to neighbouring Greece and Romania, Bulgaria remains highly exposed to supply risks related to Russian gas. As revealed by the 2014 EU stress tests, carried out against the backdrop of Russia's annexation of Crimea and Moscow's support of rebel groups in the Ukraine's eastern breakaway provinces, Bulgaria – and South Eastern Europe more generally – remains among the most vulnerable EU countries (European Commission 2014d).

What's more, Bulgaria qualifies as 'energy poor' (Bouzarovski 2013). Its citizens spend a comparably high share of their available income on energy services, with a particular problem resting on household heating and cooling (Pye et al. 2015). Given this, energy pricing has emerged as a highly contagious political issue. To be sure, energy prices in Bulgaria have remained low compared to other countries in the region, a function of heavy-handed state intervention in price setting. Yet, EU-level policies in the shape of the 20–20–20 goals suggest a pressing need for Bulgaria to reform its energy sector, notably with a view to making energy use more efficient (European Commission 2010). This will most likely imply increasing price levels to foster energy savings and incentivize additional investment into lowering the Bulgarian economy's energy intensity. Therefore, persisting inefficiencies in energy consumption coupled with low income levels present Bulgaria with a significant social challenge. In light of this, energy sector reform has stalled against the backdrop of public opposition and for fear of social hardship, while even modest price increases have repeatedly led to protest, in some cases forcing governments to leave office (The Economist 2013a).

Further, and relatedly, Bulgaria faces the challenge of replacing electricity with other fuels for heating and households. For the government, providing households with more efficient energy services represents a key means of fighting poverty and lowering energy bills (Ministry of Economics, Energy and Tourism 2011). While enhancing the role of coal – currently making up some 40 per cent of the country's energy consumption – is an option, EU climate policies put a cap on a massive expansion of solid fuels in the energy mix. Renewables, in turn, have been expanded significantly, and had already reached the agreed target of a 16 per cent share by 2012 (Eurostat 2014). However, for reasons related to overshooting costs, the Bulgarian government shies away from expanding renewable capacity further and aims at keeping its current share in the energy mix (Reuters 2015a). Moreover, although nuclear energy (20 per cent of overall consumption) makes Bulgaria a net exporter of electricity, its future is unclear. For safety reasons, two reactors at the Kozloduy power plant, the country's single facility, were put offline in 2004, and another two in 2007. Plans for capacity expansion remain in limbo, for reasons related to costs, despite public support (AFP 2015; Euractiv 2013). Bulgaria's plans to build a second plant at Belene have been abandoned due to a lack of foreign investors (Novinite 2012). This leaves gas as the fuel of choice.

Bulgaria's goal of connecting 30 per cent of the country's households to the gas grid, up from less than 2 per cent in 2013, implies a steep growth in natural gas consumption going forward, and a significant expansion of the currently small (3 bcm-volume) Bulgarian gas market. The question emerges where this gas might be sourced from, and at what price. Bulgaria, the EU's poorest country, has in the past paid among the highest gas prices in Europe, a function of its lopsided dependence on one external supplier and lack of alternatives (see, for instance, European Commission 2014k). Addressing the energy poverty challenge by way of perpetuating the country's dependence on Russia may therefore not only fall foul of security considerations but also run counter to social policy goals. In all, the policy context in Bulgaria suggests significant trade-offs between development goals, EU-level policies, international security imperatives and considerations related to household economics.

Against this backdrop, Bulgaria's shale gas emerges as an interesting energy policy option. The country's domestically available reserves, as estimated by the EIA, could numerically cover current gas consumption for roughly 185 years (see Figure 5.1) (EIA/ARI 2013). Bulgaria's former Energy Minister, Traicho Traikov suggested shale gas reserves could even last 300 years (Georgiev 2011). What's more, the country's reserves might also give rise to a domestic unconventional gas industry, which, as perceived by Bulgarian politicians, could fuel

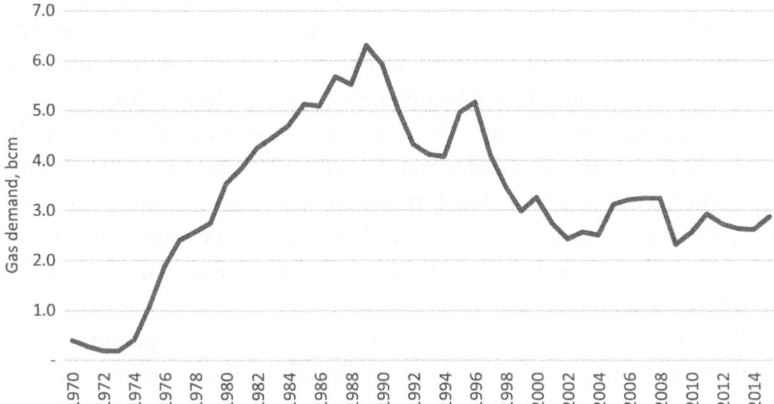

Figure 5.1 Bulgaria's gas demand, bcm
Source: BP (2016)

development and enhance supply options. Additional indigenous supplies in the shape of shale gas could, jointly with offshore gas from the Black Sea and eventually the Southern Corridor, enhance gas-on-gas competition in the Bulgarian market, putting pricing pressure on imported Gazprom gas. Consequently, the current Bulgarian energy strategy lists shale gas as an important source of supply and as a priority area for investment in exploration (Ministry of Economics, Energy and Tourism 2011). It has been the government's stated intention to foster the exploration and development of domestic unconventional energy reserves and to make shale gas an integral part of the country's energy portfolio 'in five to 10 years' (Reuters 2010).

And yet, Bulgaria enacted a ban on the hydraulic fracturing technology in 2012 (The Guardian 2012), which effectively stalled further exploration activities in unconventional gas in the country (LaBelle and Goldthau 2014b). The government also withdrew the exploration permit awarded to Chevron, the company piloting Bulgarian shale prospection, as a consequence of which the only foreign major active in Bulgarian shale gas left the country (Standard News 2014). Due to this ban, the Texas-based Park Place Energy will not further pursue shale gas prospecting (for which it holds a licence), nor will the American–Bulgarian energy company Direct Petroleum. Instead, both companies have decided to focus on conventional gas prospects (EIA 2015b; Natural Gas Europe 2012).

In short, the Bulgarian shale gas story ended before it began. To be sure, this still leaves Bulgaria with the option of connecting its gas grid to neighbouring Romania and Greece, in order to enhance diversification of supply routes, if not sources. EU law also leaves decisions on the national energy mix to member states, including the option of enhancing nuclear – a route the country may indeed take, despite the questionable economic fundamentals (AFP 2015; Euractiv 2013). That said, Bulgaria's empirical track record in fostering gas interconnector projects is poor (Euractiv 2015), and enhancing the country's nuclear capacity will not solve some of the more fundamental challenges facing its energy sector.

This policy outcome in Bulgarian shale warrants explanation, not only with a view to the domestic political context, but also against the backdrop of the stark policy divergence that exists compared to Poland. As this chapter will detail, the reasons for the failure of the Bulgarian government's shale gas policy agenda lie in the lack of a uniting policy narrative, coupled with poorly designed processes and the exclusion of key stakeholders.

5.2 Actors and Sector Governance

Bulgaria's shale gas sector is subject to the general principles governing natural gas, which are defined in the 2003 Law on Energy. This law lays out the public bodies and agencies governing the country's energy sector and the national energy system more broadly. Moreover, EU energy regulation, competition rules and environmental directives, which are discussed in more detail in Chapter 2, inform national policy-making and provide the broader framework in which national energy policy-making is embedded. The legal basis for extraction activities and revenue distribution are provided by the Underground Resources Act of 1999 and the Concessions Law (enacted in 2006).

Per the Law on Energy, the Ministry of Economy and Energy (MEE, in an earlier incarnation referred to as the Ministry of Economy, Energy and Tourism) is charged with developing the country's overall energy strategy and issuing permits for prospecting and exploration of energy resources, including natural gas. The Energy and Water Regulatory Commission (EWRC), an independent body, in turn regulates energy sector activities. The Ministry of Environment and Water (MEW) coordinates with the MEE on matters of environmental protection. In addition, the MEW oversees procedures related to environmental impact assessments, as per Bulgaria's Environment Protection Act. The Act also identifies several additional bodies exerting environmental oversight, including the Executive Agency on Environmental Protection, the Regional Inspectorate of Environment and Water (RIEW), the Basin Directorates (responsible for water management), the national park departments, the district governors and municipal authorities (in charge of local environmental policy). Moreover, other ministries may be involved in shale gas-related administrative aspects, including the Ministry of Health (notably with regard to the potential health impact of fracking substances).

This hodgepodge of agencies and ministries involved in energy (and shale gas) regulation and oversight leads to administrative overlap. As a result, the Bulgarian energy sector lacks a comprehensive and transparent governance framework (European Commission 2013e; see also Stefanov et al. 2011). That said, regulatory power in the Bulgarian energy sector clearly rests at the national level and remains highly centralized. Because it issues permits and controls tender processes, the MEE emerges the key administrative player in shale gas matters. The MEW resumes oversight over subnational executive bodies such as the RIEW. The latter is charged with implementing environmental impact assessments and other environmental aspects pertaining to energy investment, including fracking. Municipalities are not involved in negotiations over concessions and contracts, nor do they exert control over the oil and gas extraction activities. However, they do now profit from a split of mineral concession revenues between the national budget (50 per cent) and their own (50 per cent), as stipulated by the Underground Resources Act, amended in 2011. Before the amendment, they would receive 30 per cent. In addition to royalties from hydrocarbon extraction, municipalities may profit from other fees, e.g. for water use, as well as from second-order fiscal effects (KPMG 2012, 74–75; interview with MEW Press Office, 2012). Still, the central government retains a strong influence even over municipalities, because it sets the terms of fiscal transfers to municipal budgets (interview with Chairman of the Management Board of the Institute for Market Economics, 2014).

In terms of sector organization, monopoly structures prevail in Bulgaria's gas industry. The state-owned Bulgaria Energy Holding (BEH) covers the entire energy value chain from natural gas transmission and storage (overseen by BEH-owned Bulgartransgaz) to wholesale distribution (Bulgargaz) and electricity generation (NEK) (Georgiev 2016). Long-term supply contracts with Gazprom and its subsidiaries and intermediaries (Overgas Inc., Wintershall and Gasexport) hardwire Bulgargaz' dominance into the country's energy governance (KPMG 2012). Although the energy sector is significant in terms of its share of Bulgaria's overall GDP – a pattern that applies to the entire Eastern European region (Ernst & Young 2013) – domestic gas production has remained insignificant. Some production exists offshore, notably from the Galata gas field in the Black Sea, which is mainly developed by UK-based Melrose Resources plc (now Petroceltic). Total E&P Bulgaria, a consortium comprising France's Total, Austria's OMV and Spain's Repsol, has explored the Khan Asparuh offshore block, but so far without clear indication of reserves there. Chevron, the US company, left in 2012. Chevron had acquired a concession for shale gas exploration but, because the concession was revoked, it has not produced gas.

The reason for the low presence of foreign companies in Bulgaria also lies in a lagging implementation of EU pro-market policies, and in their poor application. On paper, Bulgaria has liberalized its gas (and electricity) markets. Yet, in practice, there exists no competition among energy service providers, which in the gas sector effectively remain local and regional monopolies. With this, the decade old state-centred model was perpetuated, even within the pro-market environment of the EU: a pattern that applies to many post-Communist countries in Eastern Europe more generally (IMF 2014). Because of Bulgaria's failure to live up to EU requirements in energy law, the Commission launched several infringement procedures against the country for not implementing Third Energy Package rules, and eventually sent the cases to the European Court of Justice (European Commission 2011b, 2013b, 2013d).

Bulgargaz also retains a dominant position in wholesale gas supplies and exclusively services major consumer industries such as manufacturing and the power sector. Moreover, the Bulgarian state retains heavy-handed price controls in the electricity and the gas end-consumer market, with prices being regulated by the EWRC. Clearly, this system is not sustainable, as demonstrated by the financial near-collapse of NEK, the state-run electricity provider, in 2015 (Novinite 2015). Further, the Bulgarian energy sector is notoriously prone to mismanagement and corruption, as is the country more generally (Transparency International 2014). On the one hand, this is a function of a serious lack of resources in the state administration (European Commission 2013e). On the other, the energy sector suffers from a dire lack of transparency and poor institutional procedures. For instance, the European Commission, in its 2014 country report, expressed 'concerns about the independence, professional stability and continuity of the management of the Regulator'; that is, EWRC (European Commission 2014g, 23). As summarized by some observers, Bulgaria's energy sector reveals a 'complete disregard for even basic rules of good governance' (Stefanov et al. 2011, 9).

Relevant social groups, finally, include environmental groups and business organizations such as the Bulgarian Federation of the Industrial Energy Consumers (BFIEC). Both have been vocal about shale gas. Bulgarian environmental movements – like their peers across Eastern Europe – are clearly less organized and less professional than their Western European counterparts (Fagan and Carmin 2011). While well networked, and although they have ties to international environmental organizations, they tend to be very localized in nature, as well as issue-focused (Krastanova 2012). That said, environmental groups have a history of visible activism in Bulgaria, and have demonstrated their ability to mobilize heavily, as in protests against the Krumovgrad open-pit gold mine and the Chelopech gold and copper mine (see later). In addition, special interest groups such as the Energy Independence Movement, the energy consumer association or the Bulgarian Energy and Mining Forum, aim at fostering pro-shale policy agendas or goals related to energy diversification.

In all, Bulgaria's energy governance system features a high degree of centralization, with partially overlapping competences among the MEE, the MEW and other national-level state bodies. Important additional stakeholders include the national energy holding (particularly Bulgargaz) and industry groups, while subnational players, including municipalities, enjoy little formal – that is, institutionalized – influence. They may, however, may have a stake in shale gas (and energy policy more generally) because of its fiscal benefits.

5.3 Bulgaria's Policy Approach: Top-Down Process Meets Exclusive Power Arrangement

Bulgaria's policy approach to shale gas is characterized by three elements. First, it featured a strong degree of top-down management. This may reflect the more general characteristics of the Bulgarian energy governance system, but it also entails an element of policy choice. Second, the power arrangement was exclusive, not inclusive, and was limited to a select number of stakeholders. Third, the procedural arrangement inhibited societal outreach and information flow, while the state apparatus – and particularly the public administration – was characterized by a strong lack of expertise and capacity, aggravating the shortcomings in Bulgaria's procedural approach to shale. As this section will argue, these elements led to a severe societal pushback against the government's pro-shale policy. Although shale gas generally enjoyed support among the general population – a 2012 poll suggested a 75 per cent approval rate (The Sofia Echo 2012) – the government's pro-shale policy agenda eventually failed and had to be reversed.

The development of domestic shale gas was primarily championed by the highest levels of government, notably then-Minister of Economy and Energy Traicho Traikov, and enjoyed the backing of Prime

Minister Boyko Borisov. The energy ministry aggressively pushed its stated objective to build an unconventional gas industry in the country, by giving high priority to finding foreign investors for shale gas exploration licences and to facilitating the deployment of the fracking technology. The top-down approach adopted by the government is epitomized by the fact that Energy Minister Traikov single-handedly decided to grant the country's first shale gas exploration permit to Chevron, a decision which was taken without consulting even his close advisors (interview with former Advisor to Minister of Economy and Energy and former Bulgarian Ambassador-at-Large for Energy & Climate Change, 2014). Obviously, the energy ministry legally plays a strong role in Bulgaria's energy policy and the sector's institutional set-up, while internal decision-making characterizes Bulgarian energy sector management more generally (Stefanov et al. 2014). Still, the approach taken by the ministry was heavy-handed even for Bulgarian standards.

The Bulgarian government's approach to shale gas policy was not only highly centralized (decision-making and oversight remained concentrated, with essentially a small circle within one ministry), it also came with implications for administrative procedures and practices. Rather than crafting policies and processes that would capture the specific characteristics of the fracking technology (and particularly its contested aspects), cater to its status as a nascent industry (rather than a full-fledged sector) and acknowledge the fact that it essentially constituted an imported technology that was to be embedded in a state-centred energy system, Bulgaria by default used existing regulatory fiscal and environmental frameworks to deal with shale gas (KPMG 2012). To be sure, the country has a history in coal mining and offshore gas production, which provides a legal framework for hydrocarbon extraction. Yet, the decision to pursue aggressive pro-shale policies made '[t]he process itself... a bit hasty' (interview with former Chairman of the Temporary Parliamentary Committee on the Study of Shale Gas, 2014) and led to a situation in which fracking was 'mechanically transferred' into the Bulgarian context (interview with member of Bulgarian Academy of Sciences and anti-fracking activist, 2012).

Moreover, the top-down process of pushing shale gas also raised doubts over the level of independence of the agencies carrying out environmental impact assessments, as required under EU law. In fact, as the director of RIEW Varna (under the authority of the MEW) argued, shale gas is a 'a political decision to proceed with . . . or not, and not so much a question of environmental policy' (interview, 2012). This suggests that both administrative procedures and state agencies were subordinated to the prime goal of fostering unconventional gas extraction. Although interviewees representing a pro-shale agenda made frequent reference to compulsory and independent environmental oversight, the government's approach – clearly intended to cut administrative corners – put in question the way environmental impact assessments were carried out, and the degree to which their outcome could be trusted. As a businesswoman and protest leader in the town of Dobrich therefore suggested, 'national and regional environmental authorities did not have any impact [on the result], but it was a political decision' (interview, 2012).

As a corollary, the power arrangement turned out to be highly exclusive. In fact, the government's top-down approach towards shale gas prevented the buy-in of key stakeholders or potential veto players. This is most visible when looking at Bulgaria's energy economy. For instance, the chairman of industrial energy consumer association BFIEC mourned that '[a]s a matter of fact, we weren't invited [to consultations] and we didn't demonstrate any interest' (interview, 2014). Interviews suggest that Bulgargaz, the domestic gas monopolist, was also kept out of the 'shale game', as was Bulgartransgaz, the transmission system operator (TSO). To be sure, given the company's focus on import and distribution, not production, Bulgargaz may not necessarily have been keen on entering the upstream business. That said, the domestic production of gas would clearly impact on its position as the market incumbent, both in terms of market share (as Bulgargaz serves the entire domestic market) and with regard to its ability to safeguard favourable incumbent regulatory frameworks (as new market players may push for pro-market policies). As for Bulgartransgaz, as a TSO, the company would lose income should less gas be imported from Russia, and were the Transbalkan pipeline – which Bulgartransgaz operates – to be used less for the transit of Russian gas to Turkey, Greece and Macedonia.[1]

As argued by Hiteva and Maltby (2014), Bulgargaz, Bulgartransgaz and, with them, BEH more generally must be regarded as 'para-statal organizations' that have a strong interest in maintaining the status

[1] I owe this point to Atanas Georgiev.

quo: a monopoly position on the domestic market, long-term gas import contracts with Russia and a strategic position as intermediaries between the Bulgarian state and the Russian companies Gazprom and Overgas (Hiteva and Maltby 2014, 127). Because of its pivotal role in maintaining low energy prices for households and industry, Bulgargaz enjoys de facto independence from EWRC, the regulator, and indeed from the Bulgarian state itself. As a consequence, the company has been able to influence national regulation and policy frameworks (Silve and Noël 2010), and to maintain its position as the incumbent, even if in breach of EU law (European Commission 2013a). For these reasons, it would have been politically opportune to involve BEH in the process of shale gas-related policy-making.

The government's exclusive approach to shale gas policy extends to societal stakeholders. Representatives of municipalities were not involved in planning the drilling activities related to shale gas exploration, and they did not get to discuss them in advance with national or regional state authorities. Instead, they typically learned about them by way of official announcements (interview with Member of Friends of the Earth Bulgaria, 2014). This would take the shape of 'a letter from the ministry when the company was granted a permit for exploration [stipulating] that we need to assist the company first to move their heavy vehicles for the seismic studies and then to provide it with land for the drills' (interview with Mayor of General Toshevo Municipality, 2012). With this, communication between the government and affected communities and environmental groups remained top-down. This pattern also characterized relations between energy companies and the public, including municipalities, NGOs and protest groups. Rather than seeking dialogue with civil society, corporations preferred to channel their communication through the state administration (interview with member of CEE Bankwatch Network, 2014). This particularly applies to Chevron. Although the company had a significant stake in the prospects of Bulgarian shale, it is reported to have hardly engaged in public debates, nor to provide technical information to the public (interview with former Advisor to Minister of Economy and Energy and former Bulgarian Ambassador-at-Large for Energy & Climate Change, 2014). Even against the backdrop of growing social protest, '[t]he business remained silent and waited for the government to deal with this problem' (interview with former Minister of Economy, Energy and Tourism, 2012).

As a corollary, information flows remained selective, and the quality of information available from official sources was questioned due to a lack of reputable institutional backing. As a member of CEE Bankwatch Network, an international NGO monitoring the environmental impact of large-scale investments, summarizes it, an element of 'secrecy' characterized the process by which the exploration licence was granted by the government, and the latter's approach to fracking technology more generally. In fact, municipality representatives were not informed about any specifics pertaining to the exploration activity on their administrative territory (interview with Mayor of General Toshevo Municipality, 2012).[2] Moreover, events organized by the Ministry of Economy and the Parliamentary Committee on Economy and Energy were perceived as being set up 'in a purely lobbyist fashion' (interview with Executive Director of Green Policy Institute, 2014), and '[t]hings were presented as if everything is going to be alright' (interview with member of За Земята – Friends of the Earth Bulgaria, 2014). This led to allgeations of the government having 'an agreement with Chevron without an assessment of the risks' for the population (interview with Chairman of Parliamentary Committee on Economic Policy, Energy and Tourism, 2012). Such perceptions were reinforced by the government keeping undisclosed the total number of concessions that were planned to be issued for oil and gas exploration, which, as suggested by the Co-chairman of the Green party, 'covered about a third of the country's territory, including the whole off-shore zone' (interview, 2012).

Unsurprisingly, shale gas started to become a deeply contested issue. Generally, the public's level of technical knowledge on fracking tended to be low (interview with representative of Bulgarian Academy of Sciences, 2012). In this situation, neither the government nor the industry took the effort to provide sufficient and unbiased information (interview with former Minister of Economy, Energy and Tourism, 2012). Moreover, against the backdrop of the government's selective approach to information and a generally low procedural transparency, data provided by any state source or by the industry were not perceived as independent. Citizens therefore resorted to the internet and other publicly accessible sources. *Gasland*, a US documentary, became a widely cited

[2] Formally, this can be explained by the contract being concluded between the energy company and MEE, which left local communities out of the process.

source of information on the contested fracking technology, and indeed a motivation for opposing it on local levels. Civil society organizations and NGOs became 'concerned [that companies] will put almost all [elements from] the periodic table, including radioactive ones, in the [fracking] fluid' (interview with former Chairman of the Temporary Parliamentary Committee on the Study of Shale Gas, 2014). Indeed, even representatives of the pro-shale end of the spectrum suggested that the government had failed to provide information to important stakeholders in a comprehensive and convincing way, instead presenting 'unsatisfactory explanations' (interviews with former Chairman of the Temporary Parliamentary Committee on the Study of Shale Gas, 2014; former Minister of Economy, Energy and Tourism, 2012).

In an environment of strong contestations over information and data, the country's leading research institutes proved ill placed to inject knowledge into public debates and policy process. In fact, they suffered a low reputation among all societal stakeholders. From the perspective of local communities, national research bodies were tied to the national government and alleged to be contracted by the MEE (interview with businesswoman and anti-shale activist, 2012) or even paid by lobbyists (interview with Mayor of General Toshevo Municipality, 2012). Industry representatives, in turn, expressed concerns that members of the scientific community had adopted biased positions against fracking that were not backed up by facts, but instead a function of reflexes against 'Western' technology (interview with member of Board of Directors of Oil and Gas Exploration and Production plc, 2012). This finding comes against the backdrop of a severe brain drain that has plagued the Bulgarian research community since the beginning of the transition process in the 1990s, depriving some of the national research institutes of expertise and capacity (Beleva and Kotzeva 2001; Georgieva 2004). With regard to the Bulgarian shale gas debate, the lack of trusted scientific evidence implied that there existed no opportunity for flanking a flawed institutional process with independent expertise and knowledge. As noted by a member of the Academy of Sciences, public opinion thus became biased in favour of the incumbent monopolists in the gas industry, notably Bulgargaz (interview, 2012).

State bodies, finally, proved unable to cope with the complex regulatory and policy challenge as presented by shale gas and the fracking technology. On one hand, this is a function of the centralized approach to policy-making, which assigned relevant ministries and subnational agencies the task of executing the government's agenda and left little room for the administration to adopt a more tailored approach. Authorities therefore typically resorted to 'control functions' and, rather than adopting a problem-solving approach, delegated issues once they found them outside their area of formal competence (interview with member of the Board of Directors and head of the exploration unit of Oil and Gas Exploration and Production plc, 2012). In other words, policy ownership did not emerge. Moreover, industry representatives and members of the civil society pointed to a lack of effective coordination among relevant governmental units across ministries and between governance levels, and to a lack of proper monitoring procedures. On the other hand, the state administration suffered from shortages in skilled personnel. For instance, the EU Commission reported 'serious understaffing' in the MEE, with only 40–60 of 600 employees working in the field of electricity and gas (European Commission 2013e, 8). Underlining the Commission's assessment, industry representatives hinted at a lack of expertise among ministry personnel. In the words of a Drilling Supervisor at Genting Oil & Gas, a private energy company specializing in oil and gas development and exploration, competencies in the MEE would need to be 'drastically improved' in order to live up to the shale gas policy challenge (interview, 2012). Similarly, the MEW did not dispose of skilled specialists and, according to some observers, essentially remained 'incompetent' to deal with pressing environmental concerns (interview with leader of Fracking Free Bulgaria Initiative, 2014). This pattern extended to SEWRC, the energy regulator, whose financial and human capacity was deemed 'insufficient' by the Commission to carry out its control functions effectively (European Commission 2013e, 8). This affected the ministries' and regulators' capacity to put in place efficient institutional procedures, to exert effective environmental oversight and to foster 'joined-up government'. Aggravating the situation, ministers and their advisors abstained from sourcing in expertise from academia or specialists from relevant research institutes, such as the Bulgarian Academy of Sciences, Sofia University and the University of Mining and Geology (interviews with representative of the Institute of Geology at the Bulgarian Academy of Sciences, 2012; member of Bulgarian Academy of Sciences and anti-fracking activist, 2012). (This was disputed by government officials, including by the Acting Head of Unit in the Water Management Directorate of MEW; interview, 2012.)

Overall, the exclusive and top-down approach as pursued by the Bulgarian government left key stakeholders outside formal institutional procedures, including the licensing process, and was not compensated by informal outreach, for instance in the shape of public consultations. Moreover, information became a strategic asset in the government's efforts to foster its policy agenda, while for societal groups, a lack of information was a motivation to stand up against fracking. Information was not part of a broader public deliberation process over the pros and cons of fracking. What's more, both the top-down process and the strategic use of information led to what observers labeled a 'broken communication' between the ruling political elite and society (interview with Co-chairman of the Green party, 2012). Put differently, the centralized institutional process prevented feedback from 'lower' to 'higher' levels of governance and vice versa, and effectively detached the government's policy agenda from on-the-ground realities. The top-down and non-participatory approach also put in question the legitimacy of the government's shale gas policy. The latter was not only seen as informed by motives that remained obscure to the broader public, but was perceived as 'endangering democracy' itself (interview with leader of the anti-shale gas movement and Co-chairman of the Green party, 2012). Finally, a notorious lack of administrative capacity aggravated the situation and prevented a more measured administrative approach to implementing the government's shale gas policy agenda. Procedural shortcomings and low institutional capacity put both civil society and the industry at odds with the government's policy. Not surprisingly, the initial support that shale gas enjoyed among the population quickly disappeared, to be replaced by social protest against it, while the private sector by and large abstained from supporting the government's policy objectives in shale.

5.4 Policy Narratives: 'Environmental Hazard' and 'Economic Sell-Out' Trump 'Energy Independence'

The Bulgarian government, led by the conservative GERB party under Prime Minister Boyko Borisov, promoted shale gas as of 2009, based on a strong energy security narrative. A key reference point here was the 2009 Ukraine gas crisis, which was used to highlight the policy imperative of 'energy independence' for the country. Traicho Traikov, the Energy Minister, made repeated references to energy as a national security issue (Novinite 2011a), one best met by fostering domestic (shale) gas reserves. Prominent members of the ruling GERB party backed this up with statements to the effect that '[s]hale gas is not only an industry, it's geopolitics' and that reliable energy supplies are 'extremely important for our national security' (interviews with former Member of Parliament and Chairman of the Energy Independence Movement, 2012; former Chairman of the Temporary Parliamentary Committee on the Study of Shale Gas, 2014). Such views were clearly echoed by representatives of other parties, stressing that '[e]nergy independence is beyond any doubt our main priority' (interview with Chairman of Parliamentary Committee on Economic Policy, Energy and Tourism, 2012).

Moreover, frequent reference was made by members of the Bulgarian government to the economic benefits of developing domestic shale gas in Bulgaria. The government particularly pointed to lower consumer prices, an argument that catered to the energy poverty agenda. Moreover, domestic energy production was highlighted as a boon to industrial production and as a means to hedge against price hikes, which – at that time – were caused by high crude prices that trickled down to gas through the incumbent oil indexation model (Reuters 2010). Further, cheap gas would foster the country's reindustrialization, help economic development and fix Bulgaria's trade balance, in addition to bringing concession fees and royalties (interview with former Minister of Economy, Energy and Tourism, 2012; see also Georgiev 2011). As for additional revenues, the Energy Minister cited 'tens of millions in concession fees a year' (Natural Gas Europe 2013a) that domestic shale would bring to state coffers. In all, as summarized by Traikov, shale gas therefore was about 'security, independence and lower consumer prices' (Reuters 2011).

Clearly, the national security and economic opportunity narratives resonated with specific societal constituencies, notably within the energy sector and among pro-market think tanks and more conservative political observers. For the Chairman of the BFIEC, 'there are two gas topics for Bulgaria. One is security of supply, the other is pricing' (interview, 2014). Along similar lines, representatives of economic think tanks argued that domestic energy sources would lower prices, which would attract investment and thus reduce the country's overall energy import dependence. For instance, a study commissioned by the Sofia-based Institute for Market Economics suggested that an optimistic scenario for domestic shale would lower Bulgaria's gas import ratio to 55 per cent and generate an additional GDP of 3.2 per cent or more, including economic 'knock-on effects' (KC2 2014). Its modeling supports statements by energy industry representatives pointing to significant second-order benefits from gas production, which in addition to royalties include corporate taxes, employment effects, the development of a viable service industry and structural gains for the economy (interview with member of Board of Directors of Oil and Gas Exploration and Production plc, 2012). Still, neither of the two policy frames as put forward by the government attracted broader support among Bulgaria's key stakeholders in civil society, among local communities or in the business sector. Instead, two alternative narratives proved more dominant and susceptible to mobilizing a societal cross-section of actors: the *economic sell-out* frame, and the *environmental hazard* frame.

The economic sell-out frame essentially centres on shale gas production as a way to secure private gains to the detriment of the broader society. This frame needs to be seen in the context of the post-Communist economic transition process during the 1990s, in which property rights were redistributed during several, largely uncoordinated waves of privatization. In many of the formerly Communist Eastern European countries, including Bulgaria, this process was widely regarded as unjust, and often times also illegitimate, as it created few winners but many losers and led to rapidly growing social inequality. Moreover, the privatization process left its mark on the political culture and caused lasting distrust in political and economic elites, as they were perceived both as representatives of this unjust redistribution of property rights and its beneficiaries (Pickles and Smith 1998; Stark and Bruszt 1998).

Seen through an economic sell-out lens, the intransparent and hasty process of awarding shale gas exploration licences suggests hidden motives on the part of the government. Against the backdrop of prevalent corruption in the public sector, this frame therefore essentially points to elements of state capture. The public distrust in the political class and the motives guiding political choices, as epitomized by the aforementioned statement of the Chairman of the Parliamentary Committee on Economic Policy, Energy and Tourism to the effect that the government may have had 'an agreement' with Chevron (interview, 2012), adds to the general perception that 'the money will not go to people, but to the state' (interview with Chairman of BFIEC, 2014). In this context, foreign investors – and private companies more generally – did not represent engines of growth and economic opportunity in the eyes of the public. Instead, they were regarded as vehicles to extract short-term rents, leaving the long-term economic fallout to the population. A summarized by a leading activist of the Fracking Free Bulgaria Initiative, '*investor* is a dirty word in Bulgaria because for them it's easy to bribe officials and sign contracts with virtually no obligations' (interview, 2014). As some observers hinted, 'people . . . have seen in the past how [investors] caused damages and disappeared after that' (interview with Chairman of the Parliamentary Committee on Economic Policy, Energy and Tourism, 2012). US officials expressing public support for the fracking technology in Bulgaria (Euractiv 2012c) were regarded as intervening on behalf of Chevron, a US company, and interpreted as serving foreign economic agendas, rather than the country's national cause.

Historical precedent, particularly in the extractive sector, fueled support for the economic sell-out frame. A case in point is the Krumovgrad open-pit gold mine, a project that remains controversial for its alleged environmental impact on the region and the local population. Concession fees for natural resources were considered low by international standards. At the same time, Canadian investor Dundee Precious Metals reported the project would start paying back in only three years (Novinite 2011b). This spurred strong local resistance, as human safety and environmental standards were perceived as being sacrificed for profits reaped by foreign investors (Kenarov 2011). With regard to shale gas in Bulgaria, members of environmental organizations and the Green party, among others, therefore made frequent reference to the experience in the mining sector, where governmental oversight was considered to having failed (interview with Co-chairman of the Green party, 2012).

To be sure, members of the scientific community pointed to the positive local effects shale gas extraction could have on jobs, infrastructure and economic growth, along similar lines to those promoted by the government. Yet, the prospects of economic benefit were not perceived as significant by societal stakeholders. For one, shale gas clearly was understood as a sector which required a skilled workforce. Therefore, the well-paid jobs were expected to go to foreign workers, with the local population at best being employed as 'cleaning ladies, drivers and security guards' (interviews with Co-chairman of the Green party, 2012; businesswoman and anti-shale activist, 2012). Also, local mayors did not see an economic case for shale gas extraction in their municipality. Concession fees were perceived as generating only a symbolic rent, without lasting effect (interview with Mayor of General Toshevo Municipality, 2012), while the damages caused by shale gas extraction were expected to outweigh the gains for the local population (interview with member of Bulgarian Academy of Sciences and anti-fracking activist, 2012).

In this context, the fracking technology was perceived as a serious risk to the economic base of some Bulgarian regions, particularly those heavily dependent on the agricultural sector. The agricultural industry represents an important part of the Bulgarian economy, as it makes up some 10 per cent of the country's GDP and 19 per cent of total employment (European Commission 2015a), a share that is significantly higher than the EU average of 3 per cent. A case in point in this regard is the northeastern region of Dobrudzha, Bulgaria's main grain producer, which became a major theater of the conflict emerging around Bulgarian shale. Protests related to shale gas exploration in the region were clearly connected to environmental concerns, but the latter extended to and were motivated by fears of negative impacts on the region's prime product. Finally, unlike in the Polish debate on the economic advantages shale gas would bring about, the element of a 'price hedge' did not feature prominently in the Bulgarian context. While the share of coal will need to be reduced in Bulgaria's energy mix, a function of EU decarbonization policies, additional nuclear capacity will likely not materialize and – for reasons discussed earlier – renewables will likely not be expanded beyond the 20–20–20 targets (that is, 16 per cent in the case of Bulgaria). This leaves the Bulgarian economy with considerable risk regarding the costs of carbon going forward – an issue that was not taken up in shale gas debates.

Overall, in terms of 'partisanship', the economic sell-out frame featured most strongly among representatives of the political left, the Green party and the civil society. That said, it was also a dominant frame at the level of local public administration, and clearly a policy narrative that was deemed valid among a broader cross-section of societal stakeholders, including members of business associations and the scientific community.

The second dominant policy frame on Bulgarian shale centred on environmental hazard. The main narrative in this frame related to the ecological and health risks that might emerge from the chemical substances used in the hydraulic fracturing process. Indeed, fracking fluids, and particularly the proprietary nature of their chemical components, have given rise to fierce debates between environmental groups, regulators and energy companies even in the USA, the 'motherland' of fracking (New York Times 2012b; Reuters 2014c). At the centre of these debates stand both the public's right to detailed information on the utilized liquids and the question of to what extent these may be toxic and might affect groundwater safety and the environment. In Bulgaria, public concerns became focused on water supply in the Dobrudzha region, and particularly on the important Malm Valanginian artesian aquifer stretching across the Bulgarian–Romanian border. The potential contamination of the aquifer was considered a severe risk, given that it provides fresh water to a significant part of the population, in addition to irrigating the agricultural land of the region (interviews with, among others, Co-chairman of the Green party, 2012; member of Bulgarian Academy of Sciences and anti-fracking activist, 2012).

These environmental concerns were rejected by members of the scientific community, who pointed to much larger risks stemming from, for instance, existing waste disposal sites located close to the aquifer (interviews with representatives of Bulgarian Academy of Sciences, 2014). That said, there existed clear rifts within the scientific community, with some academics also siding with environmental activists. The latter were criticized as being part of 'epistemic communities . . . naturally leaning in favour of conventional technologies', hence opposing novel technologies such as fracking (interview with member of Board of Directors of Oil and Gas Exploration and Production plc, 2012). Engineers in particular stressed the reliability of technical solutions and the fact that the geological circumstances would not suggest any risk for drinking water suppliers (interviews with Acting Head of Unit, MEW Water Management Directorate, 2012; member of Board of Directors of Oil and Gas Exploration and Production plc, 2012). Public officials even dismissed concerns surrounding groundwater safety as 'senseless theories' (former Deputy Minister of Economy, Energy and Tourism, 2012). Still, the policy narrative on environmental hazards gained further traction. In particular, the link between technological risks and water supply allowed shale gas opponents to mobilize beyond the environmental community, and to reach out even to more conservative segments of society, including farmers. A case in point was the support of the National Association of Grain Producers for anti-shale protests (Medarov 2013, 166), with the lobby group endorsing the protesters' main goals (interview with Dobrich protest leader, 2012). Thus, the policy narrative had come to encompass food safety and the country's agricultural heritage (Castle 2012), and made one of the country's main industry associations take sides against the government's shale gas policy.

In this context, the media was alleged to be biased, reporting on the risks while neglecting the opportunities (interview with member of Board of Directors of Oil and Gas Exploration and Production plc, 2012). Clearly, however – and as discussed earlier – information was never objective in the Bulgarian shale debate. To the contrary, the environmental hazard narrative resonating strongly among key societal stakeholders is in part a function of poor outreach efforts by the government and of the use of information as strategic assets in a top-down and exclusive communication process.

Overall, the environmental hazard frame was supported by a broad cross-section of societal stakeholders, which in addition to representatives of the political Left included municipalities, members of academia and even industry groups such as the grain producers' association.

5.5 Assessing Bulgaria's Policy Regime: Lacking Interest Representation, Low Policy Ownership and Divergent Frames

The Bulgarian case reveals a combination of poorly designed institutional procedures, low administrative capacity and a top-down policy processes which alienated key stakeholders. The power arrangement left out not only societal groups, NGOs and environmental movements, but also industry players and, arguably, state-owned energy companies that stood to lose should shale become a reality. In addition, the government was not able to put forward a convincing narrative in support of its policy goals. As a consequence, its framing efforts pertaining to security gains and economic benefits saw a low 'ideational uptake' among important stakeholders within Bulgarian society and the economy, and eventually had to give way to competing policy narratives that united a broad opposition against shale. Overall, and as

Table 5.1 ***Summative assessment of Bulgarian policy regime in shale gas***

	Analytical focus	Key indicators (high/low)	Value
Ideas	Valence of policy frames	• Degree of 'ideational uptake' among stakeholders	• Low
		• Congruence of dominant interpretative frames of governmental actors and societal groups	• Low
Interests	Constituency representation	• Involvement of stakeholders in the policy process	• Low
		• Inclusion of veto players in the policy process	• Low
Institutions	Communication	• Procedural outreach towards non-state actors	• Low
		• Transparency of the policy process	• Low
		• Information flow among stakeholders	• Low
	Joined-up government	• Cooperation of relevant state authorities on all policy levels	• Low
		• Policy ownership across policy levels	• Low
		• Flexibility in adjusting procedures and processes	• Low
		• Administrative quality	• Low

summarized in Table 5.1, the Bulgarian policy regime in shale gas can be qualified as weak.

Against this backdrop, it is hardly surprising that the Bulgarian government's shale gas policy agenda failed a short time after it was announced. What started out as scattered and localized anti-shale gas protests quickly gained momentum on the national level. In January 2012, a broad anti-shale movement organized demonstrations in fifteen Bulgarian cities, flanking activist efforts to directly engage with high-level political representatives in the parliament. Shortly thereafter, Bulgaria's Prime Minister Borisov initiated a parliamentary vote in favour of banning the fracking technology. While observers interpreted this move as an attempt to limit the political fallout (interview with former Bulgarian ambassador to Russia, 2014), the Prime Minister's policy U-turn came to the surprise even of close observers (interview with former Head of Energy Resources and Concessions Department, Ministry of Economy and Energy, 2014). As part of the government's new anti-fracking policy, Chevron's exploration permit was revoked, while the design of the moratorium on hydraulic fracturing technology even prevented exploration for conventional gas deposits (Reuters 2012a).

5.6 Excursus: Bulgarian Shale and the Russia Factor

In this context, it is important briefly to turn the discussion back to the external dimension of Bulgaria's fracking ban. It was repeatedly alleged that foreign money, notably from Russian sources, funded anti-fracking campaigns and local protest groups (Financial Times 2014a, 2014b). The obvious rationale of such a campaign would lie in the significant economic and strategic interest that Moscow retains in South Eastern Europe, which according to some observers has become a 'new key battleground' (Skalamera and Goldthau 2016) in the EU–Russian tug-of-war over natural gas. Domestic shale gas production would not only provide alternative, non-Russian sources for the Bulgarian market, but more importantly would undermine Gazprom's incumbent business model based on bilateral LTCs, which has for long secured its rents and dominance in the region (Belyi and Goldthau 2015).

Indeed, during interview, observers pointed to Russian involvement in Bulgarian shale affairs. To start with, political support for shale gas – at least on the national level – can be roughly described as divided between the Left (against fracking) and the Right (in favour). Coincidently, the Socialist party, and the political Left more generally, traditionally tends to be Russia-leaning, whereas the Conservatives are Western-orientated. (The ultranationalist Ataka party tends to be anti-Western, and sided with GERB in support of the fracking ban.) The Socialists taking sides with the anti-shale movement was therefore interpreted as 'support [of] the geopolitical interest of Russia in delaying and blocking the shale gas exploration in Europe and in Bulgaria' (interview with former Head of Energy Resources and Concessions Department, Ministry of Economy and Energy, 2014). With this, they became part of a larger 'political play' (interview with former Chairman of the Temporary Parliamentary Committee on the Study of Shale

Gas, 2014) in which Russian corporations allegedly served as vehicles for funding a sophisticated media campaign against shale (VOA 2015). Bulgaria's former ambassador in Russia hinted that 'the environment . . . is also a very good "business" for a lot of people' (interview, 2014), which suggests an element of foreign funding for domestic Bulgarian protest groups.

Yet, contrary to such allegations of foreign actors meddling in Bulgarian affairs, primary data gathered for this book suggest that the resistance against the government's shale gas policy had very domestic causes. In fact, environmental groups were deeply rooted in local contexts, and for a long time barely networked beyond their region. Moreover, interviews suggest strong resentments among municipal stakeholders against *any* outside intervention, whether from the national level or by international actors. This includes the Bulgarian government in Sofia as much as foreign companies such as Gazprom and Overgas. Further, observers hinted at overly intrinsic motivations behind the protests, and well-educated and predominantly 'young people who take Bulgarian environment and nature to heart' (interviews with former Advisor to Minister of Economy and Energy and former Bulgarian Ambassador-at-Large for Energy & Climate Change, 2014; former Co-chairman of the Green party, 2012).

To be sure, it cannot be ruled out that Russian money found its way into Bulgaria. Also, an already strong Russian involvement in the Bulgarian energy sector may have given Moscow or Gazprom some leverage in influencing domestic policy decisions. But the very localized protest movement, a rather broad supporter base comprising a cross-section of societal stakeholders and the complete absence of joined-up government in shale gas policy rather suggest that it is predominantly domestic factors that led to a failure of the government's policy. Externally concerted campaigns may at best have had effects on the margins.

6 *A No with Options: Romania*

6.1 Triangulating the Eastern European Shale Conundrum: The Case of Romania

In many respects, the case of Romanian exhibits patterns and characteristics similar to those seen in Poland and Bulgaria. Romania sits on sizeable shale gas reserves, which according to EIA estimates amount to some 51 trillion cubic feet (or 1.4 trillion cubic metres) of technically recoverable gas resources, most of which are onshore (EIA 2015b). Although so far unproven due to a lack of exploratory drilling, Romania's estimated unconventional gas reserves put the country at third place in Europe in terms of reserve holders, and could theoretically cover domestic gas consumption for roughly 130 years. As in Poland and Bulgaria, international energy corporations took interest in the country's shale gas prospects after initial EIA estimates were released. Several international and regional oil companies acquired prospection licences, including Chevron (USA), Sterling Resources (Canada), TransAtlantic Petroleum (Canada) and MOL (Hungary). Romgaz, the incumbent domestic gas producer, also stated its intent to join an emerging Romanian 'shale gale'. Again, as in Poland and Bulgaria (and indeed other CEE countries, such as Lithuania), the US energy giant Chevron emerged as the most prominent private foreign corporation in Romanian shale. The company acquired three licences in 2010 for prospecting gas in Dobruja, a Southern Romanian area adjacent to Bulgaria, and one licence in 2011 for Bârlad, located in Romania's eastern provinces, close to the Moldovan border (Dąborowski and Groszkowski 2012).

Shale gas exploration in Romania stalled soon after licences were awarded, and within a few years the high hopes for the country's promising reserves gave way to disenchantment among investors and shale gas proponents. With this, the country joined in a broader trend in CEE. Officially citing disappointing drilling results, Chevron returned its licences in 2015, while Sterling Resources passed its Romanian shale gas assets on to the Carlyle Group, an asset-management firm. This effectively ended Romania's unconventional gas story for the time being. As our analysis will reveal, however, it was not necessarily geology that made Chevron or Sterling leave, nor the competition from – possibly more attractive or economical – conventional gas prospects off Romania's Black Sea coast. Rather, it was a domestic policy regime that remained weak, failed to encompass crucial stakeholders, did not feature a convincing policy narrative and ultimately failed to provide an environment in which an unconventional gas industry could emerge.

Still, and in contrast to neighbouring Bulgaria, Romania abstained from enacting a ban on shale and remains legally open to shale gas exploration and related foreign investment. To be sure, the Romanian population is by and large sceptical towards fracking, and anti-shale protests erupted across the country in 2012, mainly directed against Chevron and its exploratory drilling activities. However, a temporary – and, in fact, technically informal – moratorium on shale gas exploration expired in 2012 without being renewed, and the Romanian government abstained from adopting further measures against fracking. In other words, fracking is still legal in Romania. With this, the country's policy approach to shale gas represents somewhat of a halfway house between a ban or moratorium (the Bulgarian case) and the proactive promotion of the development of unconventional energy reserves (the Polish case).

Moreover, Romania constitutes an outlier in Eastern Europe, to the effect that its domestic gas balance looks much more favourable than that of other CEE countries. In 2014, Romania produced 11.4 bcm of gas (BP 2015) and featured a 15 per cent dependence rate in gas imports (Eurogas 2014). This ratio went down to low single digits in 2015. In 2016, the country was expected to effectively stop being a net importer of gas (Transgaz 2016) and to cover national consumption with domestic output, although this ultimately did not happen given favourable prices for imported gas (Platts 2017). Although the spectre of Russian gas geopolitics does loom in Romania, the country's energy security concerns related to gas imports can be assumed to be less pronounced than in neighbouring Bulgaria or in Poland, both of which feature comparably high import ratios.

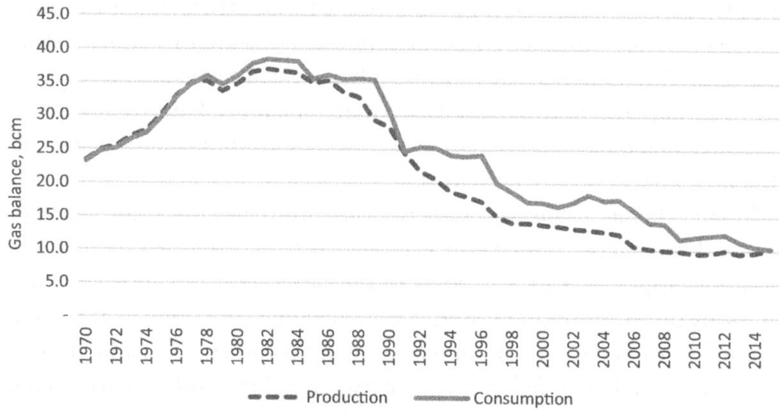

Figure 6.1 Romania's gas balance, bcm
Source: BP (2016)

That said, it is not a given that Romania will retain its currently favourable gas balance going forward (Figure 6.1). In fact, conventional production in Romania is expected to decline significantly within the next decade (KPMG 2012). Since its peak in 1982, the country's output has shrunk from 37 bcm to today's 11.4 bcm, with a reserves-to-production (R/P) ratio giving Romania another ten to fifteen years until reserves are exhausted (Departmentul Pentru Energie 2014). In other words, Romania's low import rate is not a function of dramatically increased domestic production, but rather is due to decreasing domestic gas demand over the past several years (BP 2015) – notably, in the aftermath of the 2008 crisis, which resulted in the country's industry, including its chemical and petrochemical sector, scaling down significantly, reducing its overall consumption. Other reasons include the priority assigned to electricity generated from renewable sources, the country's recent push towards deregulation of the domestic gas market (with market opening having reached more than 66 per cent of final consumption in 2015 – see Romanian Energy Regulatory Authority 2015b) and modest energy efficiency gains. Even if energy efficiency efforts are maintained or enhanced, the question of energy import dependence might therefore be back sooner rather than later, given current production trajectories. Rapidly aging gas fields will need to be replaced, notably by new offshore finds in the Black Sea. This also brings back the question of gas imports, particularly through Ukraine, which is the key transit country for Romanian gas imports from Russia. So, while Romania may sit comfortably for the time being, there exists a policy imperative to ensure the country's supply base in gas going forward. Shale gas has therefore constituted an important element in domestic energy security debates.

In sum, being a country with options, Romania does not share in some of the energy policy concerns that feature prominently in CEE, notably supply security – at least in the short term (see next section). In terms of policy output – and acknowledging that none of the three countries studied in this book has seen any commercial production so far – Romania has ended up sitting between Poland and Bulgaria to the extent that it is neither committed to shale, nor firmly opposed to its being explored going forward. With this, the Romanian case represents a handy opportunity for triangulating the empirical observations made in Poland and Bulgaria. More to the point, Romanian shale gas allows an investigation of whether – and to what extent – policy regimes play a role in countries less exposed to supply risks. As this chapter will establish, they indeed do. Deeper causes for Romania's stalling shale sector lie in the social opposition to shale gas, in problems pertaining to regulatory frameworks and institutional settings and in the fact that stakeholders were drawn from changing socio-economic constituencies, a result of the volatility characterizing the country's political system and the elite's political preferences.

6.2 Romanian Shale: Policy Context

The policy context characterizing Romanian shale gas features three key elements. First, the country's political system is highly volatile. To be sure, this statement also holds true for neighbouring Bulgaria, and for Eastern Europe more generally, particularly during the first fifteen or so years of post-Communist transition (Nikolenyi 2014). Yet, Romania represents a case of pronounced political uncertainty, as demonstrated by the country's political risk rankings, which over the years have consistently trailed behind those pertaining to Bulgaria and Poland (The PRS Group 2015). Romania's low performance in political risk rankings is a function of the country's political landscape suffering from a higher degree of political instability over the past decade compared to these other countries (which joined the EU in 2004 and 2007, respectively), particularly in the aftermath of the 2008 financial crisis. A frequently cited reason for this lies in the austerity programmes launched in 2010, which led to street riots and civil unrest (New York Times 2012a), culminating in the ousting of the conservative government headed by Prime Minister Emil Boc (Democratic Liberal Party), a short interregnum of the Mihai-Răzvan Ungureanu cabinet (Democratic Liberal Party) and the eventual rise to power of Victor Ponta (Social Democratic Party, PSD), all in 2012. Ponta moved to impeach incumbent president Traian Băsescu, and held the office of Prime Minister until 2015. He resigned in October 2015 due to mounting domestic pressure related to alleged forgery, tax evasion, money laundering and the mishandling of a Bucharest night club fire (Newsweek 2015). In early 2017, public demonstrations erupted again over the government's plans to decriminalize low-level corruption offenses, which would have primarily favoured leading PSD politicians, resulting in the resignation of the Minister of Justice. Protesters keep up the pressure on the incumbent cabinet, calling for the resignation of Prime Minister Sorin Grindeanu (PSD) and other holders of high political office (New York Times 2017). As will be discussed later, the high volatility characterizing Romania's political landscape translates into significant shifts in governmental policy preferences – both in shale gas and in policy agendas more generally.

The second element characterizing the policy context of Romanian shale gas is the geopolitics pertaining to Eastern Europe. More specifically, it is the dwindling domestic supply base in gas that has raised the spectre of Russian gas geopolitics in Romanian policy debates. As already indicated, Romania's gas balance has come to look more favourable recently, thanks mainly to declining domestic consumption and the country's starting to adhere to EU targets in renewables, market competition and energy efficiency. Against the backdrop of dwindling gas output, however, and as again already indicated, the question emerges whether Romania will be able to retain its current status of 'energy independence' in gas, or whether it might eventually go back to the times when the only source of Romanian gas imports was Russia, whose Gazprom still supplied about a third of the country's gas needs back in 2006. Moreover, as Romania represents one of the poorest countries in the EU, there exists significant upwards potential in economic growth. Growth, however, will likely spur domestic gas consumption, widening the gap between output and demand once more. To be sure, Romania still remains one of the least energy efficient nations in Europe (European Commission 2015b), and the country's significant potential here could, if utilized, make up for much of the additional demand. Going forward, however, demand sides policies will not be a panacea. By contrast, the development of the country's gas balance will also depend on the degree to which gas production rates can be reversed. In this context, offshore conventional gas prospects have moved to the centre of attention in politics and business. OMV Petrom and ExxonMobil reported the discovery of a 42–84 bcm gas deposit off the country's coasts in the Black Sea (Energyworldmag 2016). The deep-sea offshore field of Neptun could lift Romania's proven reserves between 35 and 70 per cent, and by some estimates could boost national annual production by up to 60 per cent (Georgiev 2016), even if only for ten or so years. Yet, the exploitation of Neptun remains in the planning stage, and it will require capital-intensive investment and advanced technology, neither of which has yet materialized.

Finally, Romania faces policy imperatives stemming from EU decarbonization policies and economic development, much like the ones known from Bulgaria. A key difference, however, lies in the fact that Romania finds itself on a robust economic growth path and has overcome the fallout from the 2008 economic crisis. Admittedly, this growth trajectory departs from a comparably low level, as Romania qualifies as the second-poorest country in the EU-28 (Eurostat 2015). But it has grown at more than 3 per cent in the aftermath of the financial crisis, and the Commission forecasts economic growth levels of around 4 per cent going forward, with the country's outlook being of 'above potential' (European Commission 2016c). It is important to note in this context that there also exists significant upwards potential for Romania's industry. Agriculture adds 5 per cent to national GDP and accounts for 29 per cent of the country's employment (European Commission 2016c), compared to an average 5 per cent of employment for the EU-28. Going forward, it will not only be EU climate policies putting pressure on what remains a largely inefficient Romanian energy economy. The main challenge will consist in reconciling national economic development goals with climate targets, while at the same time modernizing the country's industrial base. While Romania's energy mix has seen a growth in the share of renewables and a decline in solid fuels, natural gas still makes up for a comparably small share of electricity production. Coal remains at 26 per cent and nuclear at 18 per cent, while gas (at 11 per cent) comes in only fourth after (non-hydro) renewables (12 per cent) (Romanian Energy Regulatory Authority 2015a). Gas did gain prominence in the Romanian fuel mix in 2016, at the expense of coal. Still, there is a great likelihood that the share of gas in the power and heating sectors will need to rise steeply in order to satisfy the demand increment and to put a lid on the consumption of coal in the country.

In light of this, the Romanian government prioritizes the security of energy supply and the support of the socio-economic development in its Energy Strategy 2016–2030 (Ministerul Energiei 2016).[1] This document stresses the need to foster energy investment, and estimates overall capital needs in the energy sector in the region at some EUR 100 billion. Romania also supports EU plans for a more integrated South Eastern European gas market. Connecting the region's countries' gas grids and linking them to strategic infrastructure projects such as the planned 'Vertical Corridor', consisting of the IGB and the IBR, could significantly enhance its resilience against supply shocks. Romania could then also play an important role both as a supplier and as a transit state for gas destined for the Baumgarten hub in Vienna. And, indeed, Transgaz has started to commission works on the Bulgaria–Romania–Hungary–Austria (BRUA) interconnector, an integral part of the Vertical Corridor. Yet, given that the Trans-Anatolian Pipeline (TANAP) and the Trans-Adriatic Pipeline (TAP), TANAP's extension into the EU, prioritize Southern Europe as an export destination for Caspian gas, little is likely to feed into the Vertical Corridor any time soon (see also Chapter 2). The situation might change when the Shah Deniz 2 gas field comes online in Azerbaijan, but this will not happen before the early 2020s. The planned Bulgaria–Greece interconnector may give access to LNG landing at the shores of Greece going forward, but again there is uncertainty over the timeline. In the meantime, Romania is left with limited opportunity to benefit from additional supplies from the Southern Gas Corridor, and given the limited scale of the Southern Corridor projects as they are currently planned, the country might eventually profit only at the margins. Turkish Stream, the Russia-sponsored pipeline crossing the Black Sea, replacing the failed South Stream project, is planned as two strings of 15.75 bcm per year,

[1] At the time of writing, this document was subject to public consultation.

which would primarily supply the Turkish market by way of circumventing Ukrainian transit. Moreover, as discussed in further detail in Chapter 2, there exist significant barriers to further regional gas market integration, which have their roots in domestic politics, also in Romania. Although Romania is physically connected to most neighbouring countries and their gas grids, the existing infrastructure allows only imports, not exports. The country therefore exhibits the same energy pathologies as pertain to the South Eastern European region more generally, which include slow progress in enhancing bi-directional cross-border gas infrastructure. This perpetuates a lopsided import structure and the challenges facing Romania's gas balance going forward.

These three elements provide the background against which shale gas policy and debates happen in Romania. The next section assesses relevant actors and governance levels.

6.3 Actors and Sector Governance

Romania is a longstanding hydrocarbon producer, whose history in the oil sector goes back 150 years. By the Second World War, Romania had emerged as one of Europe's largest producers of oil (and of oil products), a reason for war-waging Nazi Germany eyeing the country's energy supplies. During Communist times, the oil and petrochemical industry was fostered as one of the country's key sectors. At present, Romania still has a sizeable oil industry by regional standards, and is the largest oil and gas producer in South Eastern Europe. The country therefore has significant experience in governing fossil fuel production and the multifaceted hydrocarbon sector. What's more, the fracturing technique has been used for well stimulation there for decades, including by incumbent Romgaz.

The main piece of legislation governing modern oil and gas E&P in Romania (lumped together under the rubric of 'petroleum') is the 2004 Petroleum Law. The country operates a standard concession system, in which E&P licence agreements are signed for up to thirty years. Upstream activities may be performed by national or foreign companies registered with domestic authorities and holding a petroleum concession for a specific area. Mid- and downstream market segments are regulated by the 2012 Electricity and Natural Gas Law. Conventional and unconventional gas are treated equally in Romania, as it has not introduced shale gas-specific legal provisions (Mihalache 2015).

Licensing and authorization procedures for conventional and non-conventional hydrocarbons are therefore identical as well. In terms of the more general principles guiding energy-sector regulation and environmental protection, it is EU laws that inform national energy policy, as detailed in Chapter 2.

Key regulatory authority in energy policy-making and regulation rests with the Ministry of Energy, which is in charge of overall sectoral oversight. The ministry implements the government's energy strategy and monitors compliance with international obligations. The National Agency for Mineral Resources (NAMR) represents the main regulatory body in the oil and gas upstream segment. NAMR is responsible for regulating all aspects pertaining to the exploration, development and production of minerals such as oil and gas. This includes organizing public auctions for upstream concessions, negotiating the terms of upstream permits, enforcing petroleum agreements and monitoring and measuring actual output. NAMR also has regulatory oversight over the country's national transport pipeline system. The Romanian Energy Regulatory Authority (ANRE), in turn, licenses companies active in the market and monitors their compliance with pertinent regulation. The National Agency for Environmental Protection (NAPM), finally, takes care of the EIA and supervises compliance with national- and EU-level environmental regulation.[2] The country's energy governance reflects the persistently centralized character of the Romanian state, which has prevailed despite administrative reforms transferring some policy competences to counties and local councils in the first decade of the 2000s.

National-level income from oil and gas extraction is mainly generated in the shape of royalties, which are set by the Finance Ministry at rates between 3 and 13.5 per cent, and by corporate tax (16 per cent), whereas municipalities and landlords benefit from duties and fees levied on drilling permits and land access. Additional second-order fees apply during authorization and permit procedures. The Romanian government is in the process of overhauling the upstream taxes and royalties system, which may result in the addition of a supplementary profit tax of 35 per cent, although this remains subject to ongoing debate at the time of writing (Pachiu, Vasiliu and Dudău 2016).

[2] For a detailed overview of Romanian energy-sector governance, see Pachiu, Mustaciosu and Dudau (2014).

In terms of sector organization, the country has welcomed energy investment by foreign companies since the end of Communism. The US company ExxonMobil has focused on exploring the country's Black Sea offshore gas reserves, while Chevron and several mid-cap companies such as Canada's Sterling Resources and the US TransAtlantic Petroleum, in addition to regional players such as Hungary's MOL, have been attracted by the country's shale prospects. Even against the backdrop of an increasing involvement of foreign companies, however, the state still retains a strong role in the energy sector. Romania's gas production essentially rests on the two incumbent domestic companies, Romgaz (state-owned) and OMV Petrom (in which the state retains a 21 per cent share since the company's privatization in 2004), which allows the government to exert continued influence over the sector. Transgaz, the gas TSO and owner of the country's national natural gas pipeline grid, is also under state control, as are Transelectrica, the electricity TSO, and Nuclearelectrica, which oversees the country's production of nuclear power. Gas imports from Russia are traditionally handled by Gazprom-owned Wintershall Erdgas Handelshaus Zug AG and Cyprus-based Imex Oil Ltd. Parties from all ends of the political spectrum regard the country's energy industry as a strategic sector, the control of which they are not ready to give up.

Turning to the social groups vested in Romanian shale, these originate from the realm of environmental protection. To be sure, Romania's environmental activism tends to be weaker than in Western Europe, as does that of the post-Communist region more broadly (Fagan and Carmin 2011). Yet, against the backdrop of environmental protection ranking low in national policy priorities and being fully institutionalized only after EU accession (Buzogány 2009), non-state groups have emerged as important societal actors in Romanian extractive industry policy. Mapping Romanian environmental NGOs active in shale gas debates, Vesalon and Cretan (2015) identify several national-level groups, including Alma-Ro and Terra Third Millennium (Terra Mileniul III) – the two most active organizations – as well as VIRA, Eco-Civica and Greenpeace Romania. Organizations such as Civil Society Initiative Group (Grupul de Iniţiativă al Societăţii Civile, GISC) in Bârlad, an area where Chevron holds exploration licences, are active on local levels. Reflecting the comparably scattered NGO scene in the country, ad hoc coalitions have started to form against the government's shale gas policy, including the Romanian Coalition for the Environment (Coaliţia pentru mediu), made up of sixty-nine separate NGOs (Vesalon and Cretan 2015, 294). Interestingly, and as discussed in more detail later, the Orthodox Church has emerged as another important 'non-governmental actor', and indeed a vocal voice in Romanian shale affairs. Thus, the country's unconventional gas conundrum involves well-organized and intrinsically conservative actors.

As per EU policies, the Romanian gas sector was opened to competition prior to the country accessing the EU in 2007, and it is gradually becoming more deregulated. Domestic energy-sector reform has proven cumbersome, however. Gas prices for industrial consumers have been deregulated, and the Romanian government estimates the size of the open gas market at 54 per cent of overall gas consumption (Departmentul Pentru Energie 2014), but gas prices for households and district heating plants remain regulated, and the government has decided to put further price liberalization on hold, despite a roadmap set up with the European Commission and the International Monetary Fund (IMF) in order to bring domestic prices in line with EU market parity by 2018 (Pachiu, Vasiliu and Dudău 2016). The Commission is hopeful that household markets will be liberalized for electricity by 2018 and for gas by 2021 (European Commission 2016c), but further developments remain uncertain. Overall, the implementation of pertinent EU policies is lacking, which prevents a competitive market regime from taking hold. A case in point is the Commission launching an investigation into the country's leading companies in the gas sector in 2017 (European Commission 2016a), targeting Romgaz, the state-owned gas producer OMV Petrom and the TSO Transgaz, and relating to alleged anti-competitive practices in natural gas exports from Romania to other EU member states. This case demonstrates that old habits die hard, as the state retains a heavy-handed role in the country's energy sector despite EU-induced privatization and deregulation. As the Commission notes in its 2014 country report, Romania will also still need to fully transpose the Third Energy Package and related legislation and to reform its state-owned enterprises in the energy sector in order to bring them in line with EU energy-market legislation (European Commission 2014h). As the next section will argue, this is not only a question of political will but also a function of the institutional approach guiding policy design and implementation in Romania more generally.

6.4 Romania's Policy Approach: Non-Participatory, Inconsistent and Devoid of Institutional Leadership

At its core, the Romanian institutional approach to unconventional gas mirrors more fundamental patterns characterizing contemporary domestic politics and governance. On one hand, shale gas policy reflects the volatile political environment in the country. Not only did shale gas policy agendas see significant swings due to changes in the country's political leadership, notably when the pro-shale Ungureanu government handed powers over to the Ponta administration, but the Ponta government itself fundamentally altered its stance towards shale gas and performed a remarkable flip-flop on fracking. In fact, Ponta partially came to power on an anti-shale platform. Rallying against the Ungureanu cabinet and President Băsescu granting shale gas concessions to Chevron, Ponta managed to overthrow the incumbent government in a motion of non-confidence in April 2012. Specifically, Ponta highlighted the environmental risks associated with shale gas exploration, Chevron's winning exploration licences over its competitors and the alleged 'secrecy' of the concessions agreement (Avocatnet.ro 2012). Following up on his campaign agenda, Ponta initiated a temporary moratorium on fracking in the country (in the shape of a declaration that was never formally enacted). However, during the general elections in late 2012, Ponta rallied on a very different policy platform, reinvented himself as a staunch supporter of shale gas and championed Chevron's presence in Romania (interview with Director of Energy Policy Group, 2014). He later stuck to a strong pro-shale position until the presidential race in 2014 (which he lost), before eventually asserting that 'Romania does not have shale gas' after all (Reuters 2014b). The government's attention subsequently shifted to fostering offshore gas exploration in the Black Sea. While the Ponta government put leadership and political capital behind developing the country's unconventional gas industry, there was clearly an element of uncertainty over its long-term political commitment. As an analyst at Expert Forum, a Bucharest-based think-tank, puts it, 'the fickle institutions probably [are] one of the most important obstacles against public acceptance of fracking. The public does not trust institutions that today want to put a moratorium on fracking and tomorrow turn into enthusiastic supporters of it' (interview, 2014).

Moreover, it is a traditional policy pattern of exclusive and non-participatory decision-making that characterizes Romania's policy approach to shale. As a corollary, the power arrangement remained limited to the highest state officials, including the Prime Minister. As revealed in interviews, Romgaz and Petrom adopted a wait-and-see approach to Romanian shale, and did not actively engage. This was partially attributed to a lack of governmental determination to get domestic incumbents involved (unlike in Black Sea offshore gas, where the Ponta government actively pushed for joint development with Exxon Mobil – see, among other, UPI 2013). Another motivation for incumbents to keep their distance from the government's shale gas policy agenda, it was suggested, lay in the avoidance of potentially bad press and the desire to 'reap the fruits when time will come' (interview with President of Terra Mileniul III, 2014). Local communities were by and large found to be ignored by the government. While this evident lack of a defined stakeholder management process might be the result of a deliberate governmental strategy, it is not inconceivable that it is also a function of the leadership's flip-flop approach to fracking, and the apparent lack of a grand strategy around related policy agendas.

This evidently non-participatory and exclusive approach to shale gas extends to the procedural arragmenet. Dąborowski and Groszkowski (2012) cite Crin Antonescu, then-leader of the National-Liberal Party, to the effect that the public did not necessarily start protests against shale per se, but against the procedures around the award of exploration licences. This points primarily to an institutional problem. At its core is the Ungureanu government's initial move to classify not only geological data, but also the terms and conditions of the concession agreement with Chevron. This sparked controversy among observers and planted the seeds of subsequent large-scale public protests. The government's refusal to disclose information on the terms and conditions of the deal – which clearly was in line with existing legislation – tied into an already prevalent public perception of secretive policy-making, and of public agencies lacking institutional integrity (Dudău 2014). In this context, observers hinted at vested interests keeping deals non-transparent. A case in point is alleged real-estate transactions with Chevron benefitting individuals holding public office (interview with President of Terra Mileniul III, 2014). Such incidents point to the larger problem of corruption, which 'persists at all levels of public administration' (European Commission 2016c, 46). As Dudău (2014) notes, the country lacks a tradition of transparency and public involvement when it comes to large-scale extractive industry projects. This goes beyond contractual details, and extends to procedural transparency. NAMR has a track record of being unresponsive to citizen inquiries and has been criticized for remaining secretive about approval processes in the mining sector, notably by NGOs such as Mining Watch (Pencea, Brădăţan and Simion 2013). In fact, as the Director of NAPM admits, '[u]nfortunately, we started off on the wrong foot, due to inadequate communication' (interview, 2014).

Against the backdrop of a non-transparent institutional process and a lack of outreach, it was access to information and the degree to which this information could be trusted that emerged as a focal point in the political discourse surrounding shale. Key aspects here surrounded fracking fluids, water contamination and environmental risks. As revealed by several experts, scientific information on fracking and unconventional energy production is hardly accessible in the Romanian language, and there exists a lack of authoritative scientific voices in the domestic public debate. While NAMR launched an information campaign on shale gas between 2013 and 2014, it dismissed by observers as insufficient and biased. Moreover, scientific studies were alleged to have been commissioned by vested interests, to the effect that 'scientists and academics that often promote shale gas are happy to get a small research contract financed by the oil companies' (interview with President of Terra Mileniul III, 2014). This led to the emergence of two 'epistemic communities' (interview with Director of Energy Policy Group, 2014), with the science and business community and the anti-shale gas community finding each other on opposite ends of the spectrum. Possibly meant as an attempt to bridge this divide, NAMR reportedly established a working group staffed by academics from national universities and tasked with mining the data on shale in Romania (Gusilov 2012), but this seems to have been the exception, and not to have had a great impact on public debates.

As a consequence, a 'facts-based' discourse was prevented from emerging, as suggested in interviews, and internet-based sources became key references points on the potential environmental and health risks pertaining to Romanian shale. A case in point, again, is *Gasland*, the US documentary, which had already risen to prominence in Poland and Bulgaria, and which started to become an important reference point in Romanian debates. Specialized Facebook groups played a role in that context, too. Moreover, it was think tanks that started to take the initiative in enhancing evidence-based public awareness, albeit with limited reach (Mihalache 2015), while NGOs and civil society tried to 'counterbalance' (interview with President of Terra Mileniul III, 2014) official policy positions. What's more, non-specialist commentators dominated public debates. During interviews, observers such as the Counselor to the Energy Minister and Member of the Management Council of Transgaz regretted the 'little reference to science-based assessments of the costs and risks that shale gas operation can bring to Romania' (interview, 2014), while others dismissed public debates and their underlying sources of information as 'scientifically unfounded' (interview with President of NAMR, 2014) or even 'hostile [and] fueled by often radical and anarchistic movements' (interview with Romanian Ambassador-at-Large for Energy Security, 2014).

Part of the explanation clearly also lies in the operative mode of public authorities in the country. Gusilov, for instance, points to a 'predominantly reactive mindset' (Gusilov 2012, 5) among the country's public administration. Along similar lines, the EU Commission stresses in its 2016 country report that '[s]trategic planning is not a common practice' in Romanian bureaucracy (European Commission 2016c, 45). The fact that information on licence agreements was not proactively disclosed – reportedly against Chevron's request (Dąborowski and Groszkowski 2012) – might therefore not necessarily be a function of a grand strategy, or of widespread corruption among the country's political and bureaucratic elite. Instead, there is an element of an institutional legacy going all the way back to pre-EU accession, and even Communist rule. The absence of a governmental communication strategy flanking pro-shale policies needs to be seen in a similar light.

Such a reading ties into another pathology in Romania: a lack of institutional quality, capacity and effectiveness. During interviews, regulatory authorities overseeing the Romanian energy sector were described as 'weak and indecisive' (interview with President of Terra Mileniul III, 2014) or 'inefficient' (interview with Counselor to the Energy Minister and Member of the Management Council of Transgaz, 2014), and were characterized by an insufficient coordination between relevant government units. This ties into the broader findings of the European Commission's 2016 country report, pointing to a limited effectiveness and efficiency of the public administration, complex administrative procedures, volatility in the tax regime and problems in human resource management (European Commission 2016c). This suggest a generally weak state apparatus governing the energy sector, as a result of which policies remained inconsistent, in terms of both design and implementation. For instance, an analyst at Expert Forum noted that the frequent and somewhat arbitrary changes in Romania's fiscal regime obviously impacted on energy investment (interview, 2014). Slow progress on the draft law on taxes and royalties, pending for years, had led to the holding back of thirty-six new concessions for on- and offshore hydrocarbon licences (Reuters 2015d). A special constructions tax, introduced in 2014, then altered the cost for rigs, pipelines and other energy-related infrastructure. It was only after interventions from the business community that amendments to this tax were introduced, in 2015, which exempted, among other things, offshore constructions particularly relevant for oil and gas E&P. Moreover, inconsistencies in putting relevant regulatory frameworks to work are likely to have resulted in similar cases seeing divergent administrative decisions, as a study of the Bucharest-based Energy Policy Group suggests (EPG 2014). Adding to that, Romania's shale gas sector operates on authorization procedures that were designed for conventional oil and gas production, leaving numerous legal blank spots pertaining to concessions, environmental oversight and tax issues. Coupled with a weak administrative capacity in shale gas issues, this leads to 'micro-management procedures' and 'ad hoc interpretations' characterizing licensing processes (Mihalache 2015).

6.5 Policy Narratives: Environmentalism and Anti-Capitalism Trump 'Economic Opportunity'

Two frames dominated Romania's shale gas debates. The first relates to the 'economic opportunity' that shale gas would bring to the country. It was a frame that essentially all Romanian governments endorsed and supported over the years. The primary narrative here centred on the potential fiscal stimulus of an emerging unconventional gas industry, and the effects on employment. For instance, Romanian President Traian Băsescu, an outspoken proponent of shale gas, called for exploring and producing the country's reserves 'to create jobs'

(Naturalgasworld 2012), whereas Prime Minister Victor Ponta – who had rallied against shale gas while still in opposition – stressed the potential benefits, especially for poor areas (Reuters 2013b). In fact, Chevron was reported to be expected to bring an estimated investment of USD 600 million into the country over the period it holds its concession (Business Magazin 2013). A study of the Romanian National Committee of World Energy Council suggests 18 069 new jobs on a national level and 19 200 on a local level by 2030. The economic stimulus was estimated at 0.5 per cent of GDP, and gas prices were estimated to decrease by 30 per cent, according to the study (Romanian National Committee of World Energy Council 2013). In addition, financially potent foreign companies such as Chevron clearly were welcome investors in the aftermath of the 2008 financial crisis, which had deep economic consequences for the country. Proponents of the 'economic opportunity' frame also pointed to the positive effects for the Romanian gas market, whose oligopoly structure would benefit from enhanced competition. As argued by an analyst at Expert Forum, a Bucharest-based think tank, '[n]ew competitors . . . would bring shale gas to the market, [which,] together with potential reserves in the Black Sea, would make competition look totally different' (interview, 2014).

Local economic development emerged as a focal point in this frame. Potential shale gas plays in the country are expected to be predominantly located in poorer, rural areas, which typically 'badly . . . need better infrastructure – water, roads, scholarships for kids etc' (interview with Director of NAPM, Ministry of Environment, Waters and Forests, 2014). As observers noted, the presence of large foreign companies like Chevron will at least temporarily have a positive impact on local businesses (Mihalache 2015). The country's shale gas prospects will therefore particularly improve 'the quality of life in the involved local communities, by creating jobs and raising local budget revenues' (interview with Romanian Ambassador-at-Large for Energy Security, 2014).

Extending the economic opportunity frame further, to the realm of national security, interviewees pointed to 'a favourable case for shale gas exploration, due to the benefits it can bring to the national economy and energy security' (interview with Romanian Ambassador-at-Large for Energy Security, 2014). As argued by the President of NAMR, shale gas was expected to foster 'energy security and stability, especially in the present geopolitical context (interview, 2014). Along similar lines,

Romanian President Traian Băsescu had earlier linked shale gas to the goal of 'energy independence' (Naturalgasworld 2013), a narrative that Ponta endorsed by suggesting that 'I'm not anti-Russian or anti-Gazprom, but it's good to be independent' (Simina 2013). Against the backdrop of dwindling domestic gas production, shale gas was seen as 'a potential substitute and alternative to conventional hydrocarbons' (interview with Member of Parliament, Industry and Services Commission, 2014), and as 'a contribution to our energy balance' (interview with Director of ANPM, 2014). Although the energy-security argument was made frequently in interviews, it still remained in the context of an overall economic welfare framing, which may reflect Romania's more comfortable position compared to other CEE nations. The 'economic opportunity' frame was most dominant among public administration representatives and independent observers.

The second – and eventually much more dominant – frame of 'anti-neoliberalism' directly opposes key assertions of the economic opportunity frame. The anti-neoliberalism narrative goes beyond a mere 'economic sell-out' frame as found in Bulgaria. In fact, it essentially represents an anti-capitalist agenda. This frame would therefore not stop at pointing to the negative economic effects of exploiting the nation's resources by private companies. Rather, it would question the system itself, as it lays the ground for this very exploitation. Importantly, through an anti-capitalist lens, the state is not a partner or an opponent in a political struggle over the right policy. Instead, it is perceived as ensuring capitalism's survival.[3] The state becomes a representative of the capitalist system itself, and hence an enemy – which, eventually, presents the Marxist with cause for revolution. This obviously also transforms the discourse over the policy agendas the state pursues, to the effect that these agendas are no longer judged on their own merit, but as part of a broader, systemic agenda.

In their efforts to 'map' anti-fracking protest groups in Romania through media content analysis, Vesalon and Cretan (2015) find evidence of such an anti-capitalist agenda. They stress the potential of such protest to challenge the 'neoliberal consensus' dominating the post-Communist era, and with it the economic policy blueprints

[3] Instrumental Marxists would suspect policy-makers of sharing a common background with business elites, which makes their decisions reflect capital interests. Structural Marxists, by contrast, essentially see the state as reproducing the logic of capitalist institutions (Gold, Lo and Wright 1975).

centring on privatization and deregulation of natural resources. The state, in this context, is portrayed 'as a collaborator with multinational corporations in the process of the commodification of nature' (Vesalon and Cretan 2015, 300) – precisely reflecting a classic Marxist reading. In his study on Romanian perceptions of shale gas, Visan (2013) also establishes such a link when showing that the Ungureanu cabinet and President Băsescu not only supported shale gas development but also personified the very austerity policies that hit ordinary people so hard in the aftermath of the 2008 financial crisis. With this, the Băsescu presidency's policy choices epitomized a broader pattern (the neoliberal agenda), and it is this pattern that protests arguably targeted, rather than a specific policy adopted in the realm of a specific sector (such as pro-shale policies in energy). Moreover, police and security forces intervening on behalf of Chevron in the village of Pungeşti in October 2013 – or, in the words of the President of Terra Mileniul III, 'the gendarmerie [being] mobilized to defend a private company' (interview, 2014) – further supported perceptions of 'the state' acting on behalf and in the interests of 'the capital'. While police forces protecting private property is hardly controversial, it is the way in which it happened that raised opposition. Police forces guarding Chevron's exploratory drilling, so it could go forward as planned, were criticized for the violence of their on-site intervention (The Guardian 2013b), which sparked nationwide solidarity and at the same time raised 'serious concerns about human rights infringement' (interview with analysts at Expert Forum, 2014).

In this context, Visan (2013) points to the broader issue of private companies being perceived negatively for exploiting Romania's natural resources. This goes back to the 'economic sell-out' frame prevalent in Bulgaria, as across CEE more generally. Key claims made as part of this frame centre on the benefits for the public being negligible compared to the vast private gains to be made by corporations. Shale gas would bring '[no] opportunities for the locals – other than few low skills jobs' (interview with analysts at Expert Forum, 2014). Permanent positions would be open to specialized employees only, likely from abroad (interview with President of Terra Mileniul III, 2014). Financial benefit for local communities would remain limited. As to the latter point, it was revealed during interviews that the problem consists in the way revenues from hydrocarbon extraction are redistributed, as only a small fraction is channeled back to municipalities, while the prevalent tax and royalty system hardly provides opportunity for direct revenue streams for local communities. Local taxes are limited to a small share of the constructions tax – which, as discussed, was overhauled to limit its impact on energy investment. Although the President of NAMR – like the representatives of other public agencies – insisted that the net effect would be positive for local communities, he admitted that this would be through redirected transfers only (interview, 2014). As a Member of the Management Council of Transgaz, Romania's TSO, argued, '[t]he fact that currently there is no participation of local communities to the prospective revenues makes them justifiably frustrated' (interview, 2014).

Yet, there are also voices turning an anti-capitalist argument into an outright imperialist one. This twist is evident in public statements made by key figures in the protests surrounding the Pungeşti events in 2013. A case in point is Father Vasile Lăiu of the Romanian Orthodox Church, a prominent shale critic who reportedly compared foreign energy companies with an army invading the country, and went on to liken corporate managers with 'enemy chieftains' and lobbyists with 'traitors' (The Guardian 2013a). The undertone here obviously extends to nationalism, which reveals an emerging link between the anti-capitalist critique and openly nationalistic sentiments in Romania's shale gas discourse. In fact, these two agendas merging is not new to Romania, and similar patterns surfaced in the context of the 2004 privatization of Romania's Petrom, of which a 51 per cent stake went to Austria's OMV, allegedly at below market price, triggering domestic debates about a sell-out of national assets to foreign investors (Gabor 2007).

The nationalist cause overlapped with an existing conservationist and environmentalist frame, which had emerged in the context of an open-pit gold-mining project at Roşia Montană in Alba county (Transylvania). Roşia Montană turned out to be strongly contested for the use of the cyanidation mining technique (The Guardian 2013c) – which, because of its toxic compounds, may cause severe hazards to wildlife and habitat if not handled properly – particularly against the backdrop of the low environmental enforcement standards bemoaned by interviewees. The effects of a potential accident were demonstrated to the Romanian public during the 2000 spill in Baia Mare, when large volumes of cyanide-contaminated water polluted the Someş River and caused lasting damage to the environment. While Baia Mare was operated by Aurul, a gold-mining company jointly held by the Australian company Esmeralda Exploration and the Romanian government, the Roşia Montană Gold Corporation was controlled by Canada-based Gabriel Resources through an 80 per cent share. Both foreign companies, and their alleged role in bribing public officials, were viewed very critically. The Roşia Montană project was eventually brought to a halt by local- and national-level protests, which led to the village being declared a 'historic site of national interest', essentially forbidding further mining activities (The Guardian 2016b).

Roşia Montană impacted on the Romanian shale gas discourse in three important ways. First, there already existed a protest movement centred on the extractive industries and their potentially damaging environmental consequences, which shale critics were able to tie into. Second, the role of foreign companies in the extractive industries had been brought to prominence in public debates, and was viewed as generally negative. Third, the public discourse surrounding Roşia Montană centred on the importance of the natural site and its Roman artifacts, bringing in the notion of Romanian cultural heritage. These three elements resonated in shale gas frames predominantly revealing an anti-neoliberal agenda, but made the latter compatible with both a nationalist and an environmentalist agenda, each of which had broad sets of constituencies behind it. Indeed, while national-level NGOs, civil society and think tanks tended to rally on the environmental platform against shale, locally active groups focused on the economic effects. In the words of one analyst, '[p]eople in Bucharest are concerned about earthquake risks or water pollution [while] local communities protest out of fear that their agricultural [economic base] will be damaged' (interview with analysts at Expert Forum, 2014). Both, he added, 'are augmenting each other'. The national cause, embodied in the involvement of the Romanian Orthodox Church, brought these two elements together together and opened protests up to broader constituencies among the population. With this, the 'national cause' ended up being defined not in terms of job creation, supply security or economic opportunity, as advocated by various Romanian governments over the years, but by preserving natural habitat and fighting the neoliberal and capitalist agenda that was portrayed as detrimental to the nation's treasures.

To be sure, reference was made in interviews to the positive effects of '[b]ringing more gas to the energy mix . . . replacing coal [which] would bring down emissions in Romania' (interview with analysts at Expert Forum, 2014), which clearly is an environmentalist argument. Moreover, interviewees pointed to the longstanding history of Romania's oil and gas industry and to the expertise in the country with regard to both sector governance and environmental oversight (interview with Member of Parliament, Industry and Services Commission, 2014). But these arguments did not resonate in public discourse or among proponents of the environmentalist cause, who rallied under the umbrella of the more dominant anti-neoliberalism frame. Interestingly, some observers pointed to even 'market liberals' sharing in the anti-shale sentiments. This clearly had nothing to do with an anti-capitalist narrative, but was because a knowledge-driven and nature-friendly economy was considered superior to one based on natural resource extraction. This would make such liberals favour the government-supporting sectors such as IT, the creative industry and tourism, and disfavour oil and gas – and unconventional hydrocarbons with them.[4]

6.6 Assessing Romania's Policy Regime: Weak Procedures, Clashing Frames and Political Volatility

As the preceding discussion demonstrates, several key elements came together in the case of Romania. First, frequent changes in the political leadership resulted in policy agendas shifting towards shale. This, arguably, not only led to confusion among the population and businesses (and indeed, disappointment among supporters particularly for Victor Ponta), but prevented the identification of key stakeholders to be included in the power arrangement and the development a consistent institutional strategy within the state apparatus. This adds to a generally weak public administration, characterized by a comparably low institutional capacity, non-participatory and opaque administrative process, and even elements of state capture. As in Bulgaria, post-Communist legacies remain strong in Romania. Coupled with an inadequate legal framework around unconventional gas, this results in both distrust among the population and inefficiencies when it comes to policy implementation.

Second, with the government not showing clear political leadership from the outset, the policy narrative behind developing shale

[4] I owe this point to Radu Dudau.

Table 6.1 ***Summative assessment of Romanian policy regime in shale gas***

	Analytical focus	Key indicators (high/low)	Value
Ideas	Valence of policy frames	• Degree of 'ideational uptake' among stakeholders	• Low
		• Congruence of dominant interpretative frames of governmental actors and societal groups	• Low
Interests	Constituency representation	• Involvement of stakeholders in the policy process	• Low
		• Inclusion of veto players in the policy process	• Low
Institutions	Communication	• Procedural outreach towards non-state actors	• Low
		• Transparency of the policy process	• Low
		• Information flow among stakeholders	• Low
	Joined-up government	• Cooperation of relevant state authorities on all policy levels	• Low
		• Policy ownership across policy levels	• Low
		• Flexibility in adjusting procedures and processes	• Low
		• Administrative quality	• Low

was not targeted, as flip-flopping policy agendas necessarily come with changing constituencies among stakeholders. Moreover, the 'economic opportunity' frame, as advanced by Băsescu, and eventually also Ponta, did not resonate with key constituencies in local communities, businesses or the Orthodox Church. Instead, the competing and eventually more dominant narrative defined the national cause as anti-neoliberal and environmentally conservative, flavoured with nationalistic undercurrents. A Romanian specificity probably also emerges from the widespread corruption pertaining to large-scale extractive projects, which fueled anti-business sentiments and undermined trust in holders of public office.

Overall, therefore, the policy regime in Romanian shale gas was weak. More to the point, compared to Bulgaria and Poland, the lack of clear political leadership failed to establish the momentum for the emergence of a comprehensive policy regime on shale. Table 6.1 summarizes the main findings.

A lack of participatory governance, the absence of a policy narrative coupled with a clear communication strategy and a rather confrontational stance towards some of the key stakeholders all gave rise to protests. Local protests – albeit ones picked up by the media – against shale gas development in Romania occurred in the towns of Bârlad and Vama Veche in 2012, followed in 2013 by nationally organized events in the nation's bigger cities of Bucharest, Timişoara and Cluj-Napoca, among others. Bucharest's Piaţa Universităţii emerged as a focal point for the protests, with weekly events organized by the anti-shale group Uniţi Salvăm (United We Save), whose members overlapped with protesters against the Roşia Montană project (interview with Director of Energy Policy Group, 2014). Facing public opposition, the government pivoted during the 2014 presidential campaign and championed offshore conventional gas development instead of shale. Given the lack of an urgent geopolitical imperative to diversify gas sources, this can be seen as a strategy to avoid continued political confrontation over shale gas while keeping the door open to later development.

7 *The Comparative Public Policy of Shale Gas in Eastern Europe*

7.1 Comparing Policy Regimes

Before drawing a line under the empirical discussion, it is worth reminding ourselves of where CEE governments started from regarding domestic shale gas: a clear economic incentive to develop indigenous shale reserves, underpinned by a security imperative in the shape of a high import dependence on Russia; a persistent post-Communist regulatory and infrastructure legacy, which extended to the pivotal role played by large incumbent (state) corporations; and – compared to the USA – a challenging socio-economic environment. Regardless, governments across the region were determined to make shale happen. In Poland and Bulgaria, shale gas policy agendas were essentially identical, and even changes in government did not affect the principle thrust of Warsaw's policy goals in unconventional energy. Romania, by contrast, constitutes something of an outlier. For most of the time span investigated in this study, Romania's political leadership did show determination and commitment to exploring unconventional gas. That said, the government flip-flopped on shale, and made U-turns in its policy agendas.

Despite largely similar policy agendas, the policy output evidently suggests high divergence in the extent to which the governments' policy initiatives were adopted and implemented. As discussed in detail in the previous chapters, Poland set in place a shale-specific regulatory regime which, despite being criticized for being 'too little, too late' in the context of a changing external gas market environment (BMI Research 2015), defined the framework for unconventional hydrocarbon extraction. In Bulgaria, the licences Chevron had acquired were revoked, and a fracking ban was put in place by Act of Parliament. In Romania, the pro-shale policy agenda was essentially put on ice, with the government, under public pressure, refocusing its attention on offshore gas reserves. As the comparative assessment in this chapter highlights, these findings coincide with a strong shale gas policy regime in Poland and weak regimes in Bulgaria and Romania. Put differently, the empirical evidence points to different degrees of regime strength correlating with policy divergence in the three countries, and explains whether pertinent frameworks were adopted and implemented successfully or not. With this, the general hypothesis as put forward in Chapter 3 is confirmed.

Comparing the policy regimes in Poland, Bulgaria and Romania in more detail, the three cases exhibit important differences regarding the power arrangement, organizational arrangement and policy paradigm. In terms of actors, the power arrangement in Polish shale gas policy comprised key stakeholders among pertinent ministries, within the energy industry (including state-owned corporations) and in society. This approach accounted for actors that were pivotal because of their crucial role in the energy sector or their position in Poland's economy more generally, or due to their potential ability to mobilize resistance. Whether the approach adopted by the Polish government represents a conscious choice is not so much the question here, and is beyond the scope this research. What matters is that the involvement of actors beyond the nascent shale gas industry ensured the inclusion of important socio-economic stakeholders in the Polish power arrangement.

This contrasts with Bulgaria, where pivotal private, public and social players were not part of the power arrangement. Left out were not only important governmental agencies but also senior governmental officials within the very ministry in charge of drafting the country's shale gas policy. Equally left out were industry associations, NGOs, environmental movements and representatives of municipalities. Strikingly, even state-owned energy monopolists such as Bulgargaz were evidently not (institutionally) involved in shaping the country's policy approach to shale. Similarly, in Romania, the power arrangement remained exclusive and limited to the highest state officials. Non-governmental groups, local municipalities and even industry players were not included in deliberating the country's shale agenda, while corporate stakeholders such as Petrom and Romgaz remained bystanders, not active stakeholders. The Romanian government's non-participatory approach to shale may be by design; it may, however, be a function of a lack of strategy, as reflected by the political leadership's readiness to perform a volte-face on fracking if it became politically more expedient. In both the Bulgarian and the Romanian cases, the actor set comprised a relatively small number of mainly high-level government players and remained exclusive, barring even potential pivot players within the economy and society.

Regarding the organizational arrangement – institutions and procedures that structure shale gas policy-making and implementation – all examined countries revealed a distinct set of similar patterns: significant post-Communist institutional legacies; a lack of coordination between relevant state authorities; and non-transparent administrative processes. Coupled with a general lack of administrative capacity and red tape, the institutional and procedural approach hardly qualifies as representing an exercise of 'joined-up government'. Neither in Poland nor in Bulgaria and Romania were relevant state agencies mobilized and capable of working together to foster the government's shale gas policy agenda on all levels. Policy ownership remained low among state bureaucracies. This ties into public agencies being reactive rather than proactive on shale-related issues ranging from regulatory processes to information sharing. Moreover, rather than syncing policies across administrative units and communicating them effectively, ad hoc procedures prevailed and processes remained opaque for outsiders.

In Bulgaria and Romania, there also existed few entry points for voicing concerns or influencing the policy design, notably for stakeholders outside the power arrangement. Overall, institutions and procedures pertaining to shale gas were not part of a grand design and did not live up to the ideal of 'joined-up government'. Regulatory inadequacies such as frameworks designed for the incumbent conventional gas sector tended to persists, buy-in opportunity for socio-economic stakeholders was kept low and information sharing on siting decisions, the environmental impact of unconventional exploratory drilling and pertinent processes remained the exception rather than the rule. This gave rise to or cemented widespread distrust among the population and other stakeholders in the shale gas policy agenda as fostered by the different governments, delegitimizing both their policy goals and the process.

The Polish case is different on two accounts. First, unlike in Bulgaria and Romania, the policy agenda was not imposed on stakeholders in a top-down manner. Instead, and mirroring a much more inclusive power arrangement compared to the other cases, state authorities opened venues and procedures to reach out to important stakeholders on national and subnational levels, ensuring their voice was heard and buy-in could happen. Notably, when co-opting national energy companies into the process, the government turned pivotal incumbent players into stakeholders in the shale agenda. This contrasts with the Bulgarian case, where a top-down approach added to poor institutional procedures and a narrow power arrangement, resulting in the alienation of key stakeholders. It also contrasts with Romania, a case exhibiting an equally exclusive and overall non-participatory approach to decision-making in shale gas policy.

Second, local-level formats of community engagement, although often quasi-experimentalist, ad hoc and non-formalized, demonstrated room for institutional flexibility, and there was indeed willingness among public officials to explore this flexibility. To be sure, the Polish approach to shale gas policy was not one of grand design either, and the organizational arrangement featured pathologies like those of Bulgaria and Romania. This pertains notably to a general 'misfit' between the government's stated policy agendas and the ability to pursue them against the backdrop of limited administrative capacity, a lack of coordination and a regulatory regime designed around the conventional gas sector. A more outreach-focused approach, coupled with institutional flexibility, however, facilitated information flows and the inclusion of municipality-level actors, and to some extent compensated for the institutional shortcomings characterizing the Polish setting. This contrasts with both the Bulgarian and the Romanian cases, where a history in corrupt practices surrounding large-scale mining projects provided the background for highly negative preconceptions on the (local) politics of the extractive industries. Intransparent processes regarding the licensing and siting of exploration projects therefore reinforced pre-existing low levels of institutional trust among local constituencies.

As for the policy frames, all three cases share similar narratives on shale gas, albeit with a slightly varying policy focus and varying degrees of uptake among key constituencies. The 'national security' frame is common to all cases. While generally highlighting the imperative of energy independence against the backdrop of Russian energy geopolitics, it was Poland's history and national integrity that served as a primary reference point for the government's pro-shale narrative. In Bulgaria, by contrast, it was insecurities stemming from gas transit. In Romania, national security was framed in terms of retaining a stable national gas balance going forward and of preserving the country's presently favourable energy situation. A second frame, centring on 'economic opportunity', refers to job creation, competitive advantages for domestic industry thanks to lower prices, additional state revenues and knock-on effects for local development. While this narrative is common to all three cases, in the case of Poland it also promotes shale as an insurance policy against EU decarbonization targets. The Bulgarian and Romanian cases do not prominently feature this element.

Both the national security and the economic opportunity frames, as promoted by the government, resonated strongly with core constituencies in Poland, and enjoyed a high uptake in the business community and the state apparatus, as well as within society more broadly. Uptake was even found among some parts of the environmental community and some NGOs. With economics and security becoming the dominant frames around fracking, shale rose to what amounted to a national project. In Bulgaria and Romania, by contrast, the uptake remained limited beyond special-interest constituencies within parts of the energy sector, market-liberal think tanks and the foreign policy community. In Bulgaria's civil society, local communities and even parts of business, these frames did not resonate. Uptake remained similarly low among the bulk of socio-economic stakeholders in Romania. In addition to national security and economic opportunity evidently constituting weak frames, this arguably is a function of a lack of a consistent policy agenda on the part of the Romanian government, making the policy narrative less targeted.

In addition to the policy narratives as promoted by national governments, several rival frames emerged in the country cases studied. A narrative that featured prominently in Bulgaria and Romania, and only at the margins in Poland, relates to 'environmental hazard'. To be sure, a debate on the environmental consequences and risks of fracking also emerged in Poland. Yet, no dominant alternative narrative emerged from that debate. In Bulgaria, by contrast, technological risks and the potential impact on an important acquirer rose to prominence as a key theme, and as a dominant policy narrative surrounding shale. Similarly, in Romania, the fracking technology was framed predominantly in terms of environmental hazard, particularly in the context of the extractive industries and experience in gold-mining projects there. Risks for the habitat, the spectre of toxic spills and concerns related to groundwater safety all culminated in that narrative.

Another dominant frame in Bulgaria and Romania centred on foreign investment in domestic shale as a case of national 'economic sell-out', suggesting that shale gas licences and eventual production primarily served the economic interests of foreign actors, namely international investors and energy companies. While the environmental hazard narrative can be regarded as an alternative to the government's policy frame, the economic sell-out frame represents a clear counter-narrative to the government's 'opportunity' frame.

Both frames are related, and they share overlapping constituencies in terms of 'ideational uptake'. In Bulgaria, a cross-section of societal stakeholders shared in both frames, including municipal-level politicians, national business associations, members of the scientific community and representatives of the political Left. The environmental hazard frame allowed an extension of the notion of risk to food supply, thus capturing the support of conservative stakeholders such as farmers, or even the national grain producers' association. Exhibiting both anti-capitalist and nationalist undertones, the Romanian case featured a specific version of the 'economic sell-out' frame, which allowed socially conservative constituencies such as the Orthodox Church and societal groups critical towards foreign investors to be united behind a common cause. As discussed, the open-pit gold-mining project at Roşia Montană served as a focal point – and, indeed, galvanizing element – for linking the anti-capitalist and nationalist frames back to the environmental hazard frame, ensuring a high uptake among a variety of different social groups and stakeholders, from the national to the local level.

These findings generally confirm the specific hypotheses put forward in Chapter 3. The presence of a comprehensive power arrangement involving important economic and societal constituencies generated support for governmental shale gas policy agendas. Accounting for stakeholder interest facilitated the adoption of these agendas in the shape of pertinent regulatory frameworks and enabled governments to push forward the exploration of unconventional energy. In the absence of a well-shaped power arrangement, governmental shale gas policies, by contrast, saw little support and were met with resistance. Shale gas policy agendas also clearly succeeded when underpinned by a convincing policy idea or narrative. Interpretative frames exhibiting a high degree of 'valence' for core constituencies became dominant and helped ensure stakeholder support. Absent a strong and commonly shared policy paradigm, alternative frames or counter-narratives emerged, challenging the government's policy agenda. With this, SH1 and SH3 are fully validated.

Table 7.1 *Summative assessment of Eastern European policy regime in shale gas*

	Poland	Bulgaria	Romania
Ideas	• Strong policy narratives: security and economy • High ideational uptake among key stakeholders • Congruence of dominant governmental and societal interpretative frames	• Low ideational uptake of governmental policy narrative among stakeholders • Existence of dominant alternative or counter-narratives: environmental hazard, economic sell-out and food security • Strong divergence of interpretative frames	• Low ideational uptake of governmental policy narrative • Existence of dominant counter-narratives: anti-neoliberalism and (nationalist) environmentalism • Strong divergence of interpretative frames
Interests	• Stakeholder involvement/ inclusion on all governance levels	• Key stakeholders excluded	• Key stakeholders excluded
Institutions	• Ill-designed regulatory regime, low policy coordination • High policy ownership among public-sector officials • Low institutional and procedural transparency • (Informal) outreach to social interest groups (e.g. town hall meetings) • Institutional and flexible • Emphasis on information sharing	• Centralized and top-down process • Poor institutional quality, low administrative capacity • No policy ownership among public-sector officials • No institutional and procedural transparency • No outreach to potential veto players	• Non-participatory process • Weak public administration, low institutional capacity • No policy ownership among public sector officials • No institutional and procedural transparency • No outreach to potential veto players • Shifting policy agendas, preventing a consistent institutional approach
Regime strength	• High	• Low	• Low

When it comes to institutional structures and procedures, SH2 is not fully confirmed. None of the studied countries featured procedural arrangements enabling policy coordination between and across governance levels, helping policy ownership among involved administrative units or facilitating outreach to socio-economic stakeholders. Procedural arrangements remained top-down rather than allowing for participatory deliberation on the opportunities and risks of fracking, and were ill-designed to ensure information sharing with societal constituencies. They were, therefore, not susceptible to lending authority to the power arrangement or legitimacy to the policy agenda more broadly. Deviating from this assessment is Poland, where informal procedures facilitated stakeholder participation on the local level and proved flexible enough to overcome formal institutional rigidities. Table 7.1 summarizes the main findings.

7.2 Assessing the Social Licence

In addition to confirming the importance of policy regimes for explaining policy divergence, this study offers important insights into the creation of a 'social licence to frack'. As detailed in Chapter 3, such a social licence can be considered in place if countries embrace the fracking technology; if they reject it, the social licence is absent. In the case of Poland, there arguably existed and still exists a comprehensive social licence for shale gas. Precisely the opposite conclusion can be drawn for Bulgaria and Romania, where, going by the core criteria pertaining to the social licence, it was absent. The striking policy divergence in Eastern Europe therefore allows conclusions to be drawn on both the nature of the social licence and the emergence of a 'social contract' underpinning shale gas policy in the region. More to the point, the findings suggest three distinct, although intertwined, elements pertaining to the social licence to frack.

First, the social licence in unconventional gas is multilevel. It encompasses the local level, where initial test drillings happen and where the impact of exploring for unconventional gas is most tangible – the traditional focal point of the SLO literature. It also encompasses the federal (and, if policy-relevant, the regional) level, however, where broader societal consensus needs to emerge around whether shale gas is desirable for the nation in the first place; whether the anticipated benefits for the country outweigh the potential costs; and how local constituencies should be compensated for creating a national public good – energy security and welfare gains – for instance in the shape of split revenues in the royalty regime. Social acceptance, to stay in the SLO jargon, is therefore a matter of all levels of governance when it comes to fracking. In Poland, shale gas emerged as a national project thanks to a powerful and shared vision of the role the domestic resource could play in fostering the country's sovereignty and economic prospects, but also because of the economic opportunities anticipated by municipal-level actors. In Bulgaria and Romania, by contrast, national-level contestations over the role that unconventional gas could and should play in the countries' energy systems and broader economic policies coupled with local-level contestations over their say on the potential economic and environmental impacts. Both levels proved relevant in that context, and by extension in shaping the social licence.

Second, the 'social contract' in shale gas matters primarily relates to an (informal) agreement between core social constituencies and the government on an important policy concern: the socio-economic prospects and consequences of unconventional energy production. Because fracking is socially contested, and due to the complex domestic political economy behind energy policy decisions, the social contract does not manifest itself as a result of citizens simply accepting the generally agreed principle of general and fair elections and the rule of law. Put differently, the social contract is not brought about by a vote of parliament on a governmental policy proposal, or by the government acting on the basis of an existent electoral mandate. Instead, the generation of a social contract on shale gas warrants and necessitates a both specific and broad process – in addition to the following of democratic principles. It is this informal social contract that gives rise to the social licence. In the examples of Bulgaria and Romania, the (pro-shale) policy as adopted by the government was rejected by key stakeholders, who united in opposition to a – formally legal – governmental decision. This, arguably, mirrors the absence of a social contract involving core constituencies. In Poland, by contrast, the government's outreach to relevant actor sets fostered a shared and mutual agreement on the socio-economic prospects and consequences of shale, independently from formally legal aspects.

Third, the social licence is about commonly agreed principles and processes underpinning shale gas policy. It results from trusted procedures and practices, shared normative principles and policy goals, and legitimate frameworks governing unconventional gas as a contested policy issue. This aspect also partially touches upon material aspects, and relates to a fair distribution of the costs and benefits of what is considered a risky technology. With this, the social licence determines whether the governmental policy agenda is socially accepted, whether it is institutionalized in the shape of formal law and to what extent key stakeholders on all governance levels facilitate its implementation when confronted with the tangible outcomes of that policy. As the case of Poland shows, the government's policy agenda rested on a broad agreement concerning the principles of Polish energy policy, it gained the approval of core stakeholders on all societal levels, its goals were considered legitimate and credible and, although clearly not optimal, its procedures were by and large trusted. As a result, governmental policy agendas were formalized and societal actors bought into both the policy goals and the broader pro-shale agenda. In Bulgaria and Romania, by contrast, deep-seated distrust in procedures and processes coupled with contestations over the role that unconventional gas could and should play in the country's energy system and broader economic policy. Pertinent frameworks and procedures did not enjoy the necessary legitimacy among core constituencies, while cost–benefit analyses of local communities were diametrically opposed to the ones promoted by the national government. In short, there existed no agreement over the principles and processes of shale gas, and social acceptance of the fracking technology remained low on all levels. As a result, shale gas policies failed to move forward, and the contested technology remained socially contested, and in fact a politically toxic issue.[1]

Overall, for the case of Eastern Europe – a region with potentially large reserves but no significant shale gas production to date – the social licence is not so much about the physical *operation* of carrying out frack jobs in the local context (the focus of most of the SLO literature to date) but relates to some more fundamental questions of social desirability and the social acceptance of a given policy agenda that might eventually materialize in the shape of domestic gas supply. The question then emerges of what explains the creation of a social licence in the case of Poland and its absence in the case of Bulgaria and

[1] In this context, an analyst from Romania's Expert Forum pointedly suggested that 'If I were the energy minister, I'd bring in some Polish experience here. They have developed their legislation so as to reach a compromise acceptable to as many people as possible' (interview, 2014).

Poland. As posited in Chapter 3, the policy regime concept again proves helpful here. Policy regimes, by way of aligning actor interests, institutions and policy narratives, can generate a shared vision on a given policy issue among key stakeholders, engage core social constituencies towards a common goal and provide for procedural arrangements in support of both the joint policy vision and its eventual implementation. Eastern European national policy regimes comprising a well-shaped and comprehensive power arrangement, a shared policy paradigm and procedural arrangement ensuring due process seem to correlate with the presence of a social licence to frack. Weak regimes correlate with countries devoid of such a social licence. Although the obvious limitation to further generalization lies in the small number of case studies investigated in this book, the findings still suggest that strong shale gas policy regimes are susceptible to generating the necessary societal support around a shared policy agenda and the underlying processes. The causal logic is similar to that underpinning policy divergence: strong policy regimes converge the expectations of different stakeholders, ensure transparency and information sharing, and facilitate inclusive (rather than exclusive) decision-making; weak regimes do the opposite.

In addition to establishing an empirical link between strong policy regimes and the existence of a social licence, however, these findings also point to some more conceptual cues regarding how to analytically substantiate such a licence's creation. As hinted earlier, the term 'social licence' not only remains somewhat elusive but seems to lack a robust analytical foundation that would allow theorization beyond the notions of participatory governance, 'responsible minerals' and best practice (see Giurco et al. 2014; House 2013). As the case of Eastern Europe demonstrates, shale gas policy is contingent on a complex interplay of a number of factors, including highly subjective perceptions, objective material incentives and national discourses. Bringing these insights into investigations on the social licence promises to enrich pertinent works and move their focus from prescription towards explanation. The policy regime concept, originating in policy studies, may well travel therefore to the field of energy and mining policy and inform pertinent debates on the extractive industries.

8 *Conclusion*

Shale Gas, Technology Transfer and Energy Security

This book started from an obvious puzzle: why, despite comparable circumstances, did some Eastern European countries embrace shale gas as a domestic resource, while others gave it the cold shoulder? The bulk of Eastern European nations face the challenge of high Russian gas imports, have dealt with oil, gas and mining businesses in the past, are subject to EU environmental and energy regulation, are exposed to similar gas market dynamics and stand to gain from a domestic extractive industry in terms of national income and employment. And yet, Poland enacted pro-shale gas policies, whereas Bulgaria banned fracking, while Romania represents a halfway house between the two. This happened despite the political leadership being determined to push the pro-shale agenda in both Poland and Bulgaria (and mostly in Romania), with the different governments' political orientations making no difference. It also happened despite local Green movements being weaker than in Western Europe.

As our investigation revealed, the reasons for this obvious policy divergence lie in procedures and processes, stakeholder inclusion and the question whether a strong idea underpins governmental policies. Put differently, it makes a difference who is involved and how, and based on which narrative. Against the backdrop of a strong policy regime, Poland remained supportive of shale regarding both pertinent policies and social attitude. Weak regimes led to a fracking ban in Bulgaria and to shale policy initiatives effectively dropping off the agenda in Romania. With this, the empirical investigation confirmed the book's main assertions and shed light on what it takes to create a social licence for contested technologies and related policies. More generally, however, the book highlights the importance of the domestic context for understanding policy dynamics. More to the point, there are three main takeaways from the Eastern European shale gas conundrum.

First, this analysis suggests that energy security in Eastern Europe cannot be approached as a matter of 'objective' policy imperatives.

A lopsided dependence on one single supplier may indeed – all else held constant – call for policies aimed at fostering diversification of sources and supply routes, enhancing the resilience of the energy system and readjusting market structures. As discussed in Chapter 2, a plethora of works in the realm of international security and geopolitics have produced pertinent policy prescriptions to that effect. Yet, as much as there exists policy divergence in shale gas matters, Eastern European countries are also very different when it comes to pursuing these 'objective' goals, i.e. putting in place cross-border infrastructure, integrating gas markets and embracing competitive pricing regimes. Thus, their individual energy security situations differ considerably despite ongoing EU-level efforts. This drives home the point that rather than representing a rational answer to objective policy imperatives, energy policy is a function of complex domestic dynamics pertaining to institutional procedures and processes, and to winners and losers. This investigation lit on the importance of these dynamics for shaping energy policies in Eastern Europe and beyond. It calls for embedding international energy security debates in domestic policy contexts, both when it comes to explaining energy policy decisions and outcomes and in order to generate tangible and feasible recommendations going forward.

Second, technology transfer does not happen in a vacuum. As this book has highlighted, a technology that's matured in one regulatory and political environment may find it difficult to take hold in another. Domestic incumbents may have little interest in enhanced competition, regulatory regimes may prove ill-designed for a novel technology and such a technology may become contested in the policy discourse not as a result of its own properties, but for reasons contingent on the political economy context in the receiving country. The case of Eastern European shale gas provides for ample evidence in that regard. Fracking became contested in some countries not due to concerns over environment impact, but because it was perceived as serving the interests of global business over national or local welfare. Unless made part of the potential revenue stream or given the opportunity to benefit from technology transfer, state-owned companies, tied into LTCs with Gazprom and other external gas suppliers, had little incentive to give up market share and welcome the domestic competition arising from shale gas. Regulatory frameworks and practices informed by the public utility model and set in place to manage the operations of large energy utilities were not well placed to serve the needs of a nascent unconventional gas industry. Besides highlighting the domestic political economy behind technology transfer, the case of Eastern European shale gas also demonstrates how closely energy technology mutually interacts with political, economic and social forces. A decade-old co-evolutionary process between the conventional gas sector and political-economic institutions arguably shaped both, and gave rise to path dependencies that are hard to overcome. For novel technologies, such as fracking, this creates a difficult environment in which for a niche to experiment and grow. Studies on international technology transfer may profit from paying attention to the incumbent environment of the receiving country and the challenges and opportunities this environment may offer for technology adoption.

Third, policy regimes matter when it comes to socially contested technologies. As this investigation suggests, policy regimes may help establish legitimacy for the policy goals surrounding such technologies, as well as the underlying policy process, eventually creating what has been termed a 'social licence'. For the case of fracking, this implies that unconventional resource endowments may be turned into available reserves if the domestic policy regime aligns interests, institutions and ideas. The findings of this book, however, travel beyond the energy realm. In essence, they are relevant for any policy issue coming with some or all of the characteristics of Eastern European shale: a 'messy' problem cutting across various policy subsystems and governance levels; low probability but high impact risk; tangible short-term costs for a few well-defined constituencies, but unclear long-term benefits; and the prospect of winners and losers arising from a change in the status quo. With this, these findings are applicable to cases as different as large-scale physical infrastructure projects (from interstate transportation networks to cross-country electricity transmission lines), new research frontiers (such as artificial intelligence or geoengineering) and determined decarbonization pathways (requiring a fundamental rethink of national and global economic systems).[1] All of these issues are contested, and arguably warrant a social licence in order to be broadly acceptable within society. In other words, our findings suggest that the

[1] For a discussion of key aspects, including stranded assets and a looming North–South divide, see Goldthau (2017) and Manley, Cust and Cecchinato (2017).

comparative public policy of contested technologies extends to contested *policies*, and promise to open interesting avenues of research there.

To conclude, the findings of this book suggest that the deployment of new energy technologies from one country to another is a matter of many factors. Resource endowment is but one such factor. Particularly if perceived as risky, novel technologies such as fracking may end up facing the harsh reality of mounting social contestation, adding to an often unfavourable domestic political economy and a political elite that may not be prepared to deal with the policy complexities arising in this context. In addition, prices need to be right in order to set the correct incentives. Shale gas 'going global', in short, may be a long shot. It is not inconceivable that fracking will see less resistance in countries able to set aside societal concerns and to ignore the potential losers of the introduction of the novel technology. This makes China a potential non-OECD front runner, and also Russia, to the extent that the Western sanctions regime allows access to the fracking technology. In democratic contexts, however, the social licence is a prerequisite. The unconventional energy industry is therefore likely to centre on North America for some time to come. Fracking will continue to exert a global impact, even if by and large restricted to North America. It has already changed international market dynamics, and will continue to do so by adding surplus US LNG capacity to global supply going forward. Indirectly, the US shale industry has therefore already changed the energy security situation of Eastern Europe. What it might take to make the most of the new gas world is a different matter, however, and probably the subject of a book of its own.

Appendix

Table A.1 *List of country-level interviews in Poland, Bulgaria and Romania*

Position	Organization	Year
Poland		
Executive director	Cleantech Poland	2012
Project coordinator	Climate Coalition	2013
President of the board	Environmental Protection League, Lublin	2013
Researcher and policy analyst	Instytut na rzecz Ekorozwoju/Polish Institute for Sustainable Development, Projekt LIFE+ 'Dobry Klimat dla Powiatów'	2013
Advisor and geologist	Instytut Studiow Energetycznych/ Institute of Energy Studies	2012
Professor	John Paul II Catholic University of Lublin (KUL)	2013
Former advisor to Foreign Minister	Ministry of Foreign Affairs	2012
Chief expert	Ministry of Foreign Affairs, Department of Economic Policy	2012
Expert	Ministry of Foreign Affairs, Department of Economic Policy	2012
Expert	Ministry of Foreign Affairs, Department of Economic Policy	2012
Advisor to minister and to Chief National Geologist	Ministry of the Environment	2012
Department of Strategic Projects, Office of the Minister	Ministry of the Treasury	2012
Poland representative	NaftaGaz Poland	2013

Table A.1 *(cont.)*

Position	Organization	Year
Chief economist, Strategy and PPM Department	PKN Orlen	2012
Chairman	Poland's Greens 2004 party	2013
Director	Polish Confederation of Private Employers Lewiatan (PKPP Lewiatan), Department of Energy and Climate Change	2012
Expert	Polish Confederation of Private Employers Lewiatan (PKPP Lewiatan), Department of Energy and Climate Change	2012
President	Polish Ecological Club (PKE) (Central-Eastern Region)	2013
Professor	Polish Institute of Soil Science	2013
Deputy director	Regional Directorate for Environmental Protection (RDOS)	2013
Corporate affairs manager	Talisman Energy	2012
General manager	Talisman Energy Polska Sp.	2012
Professor in power engineering	Technical University of Lodz, Poland, Institute of Electric Power Engineering	2013
Country manager, business development	United Oilfield Services	2012
Professor	Warsaw University, Economics Department	2013
Bulgaria		
Anti-fracking leader/co-chair	Anti-shale gas coalition/the Green party	2012
Regional leader in anti-fracking organization	Anti-shale movement Dobrich	2012
Former member of parliament/chair	Bulgarian National Assembly/Temporary Committee on the Study of Shale Gas, GERB	2014

(cont.)

Table A.1 *(cont.)*

Position	Organization	Year
Member of parliament	Bulgarian National Assembly; chairman of the Parliamentary Committee on Economic Policy, Energy and Tourism; member of Union of Democratic Forces Party	2012
Member of parliament/ former deputy minister	Bulgarian National Assembly/Former Minister of Economy, Energy and Tourism, Citizens for European Development of Bulgaria (GERB) party	2012
Professor (retired), hydrologist, anti-shale gas activist	Bulgarian Academy of Sciences/ anti-shale gas group	2012
Professor	Bulgarian Academy of Sciences, Institute of Geology	2012
Deputy director	Bulgarian Academy of Sciences, Institute of Geology	2014
Chairman/director of strategic planning and investments	Bulgarian Federation of the Industrial Energy Consumers (BFIEC)/Stomana Industry	2014
Analyst, economic program	Center for Study of Democracy	2014
Analyst, economic program	Center for Study of Democracy	2014
Municipal deputy mayor	Dobrich rural municipality	2012
Chief environmental expert	Dobrich rural municipality	2012
Head of environmental unit	Dobrich urban municipality	2012
Mayor	Dobrich urban municipality, GERB	2012
Leader, former member of parliament	Energy independence movement	2012
Campaigner	Fracking Free Bulgaria Initiative	2014
Member	'За Земята – Friends of the Earth Bulgaria' and CEE Bankwatch Network	2014
Member	'За Земята – Friends of the Earth Bulgaria'	2014

Table A.1 (cont.)

Position	Organization	Year
Mayor	General Toshevo municipality, member of Bulgarian Socialist Party (BSP)	2012
Drilling supervisor	Genting Oil & Gas	2012
Member	Green Policy Institute, Bulgarian Greens (Political Party)	2014
Managing partner/ former ambassador of Bulgarian in Russia	Innovative Energy Solutions/Ministry of Foreign Affairs	2014
Former head of 'Energy Resources and Concessions'/energy expert	Ministry of Economy and Energy/ Bulgarian Energy and Mining Forum	2014
Former advisor to Minister of Economy and Energy and former Ambassador-at-Large for Energy and Climate Change	Ministry of Economy and Energy	2014
Acting head of unit	Ministry of Energy and Water, Water Management Directorate	2012
Member of the board of directors and head of the Exploration Unit Director	Oil and Gas Exploration and Production plc.	2012
Director	Regional Inspectorate of Environment and Water (RIEW) Varna	2012
Professor	Sofia University, Faculty of Economics and Business Administration	2012
Professor	Sofia University, Department of Geology and Geography	2012
Romania		
Director	Energy Policy Group, a Bucharest-based think tank	2014
Counselor to Energy Minister	Ministry of Energy	2014
Ambassador-at-Large for Energy Security	Ministry of Foreign Affairs	2014

(*cont.*)

Table A.1 *(cont.)*

Position	Organization	Year
Analyst, energy & public policy	Expert Forum	2014
Analyst, energy & public policy	Expert Forum	2014
President	Romanian Agency for Mineral Resources	2014
Director	Romanian Agency for Mineral Resources	2014
Member of parliament/commission member	Romanian Parliament/Commission on Industry and Services	2014
President	Terra Milleniul III	2014

References

Aalto, Pami. 2007. *The EU-Russian Energy Dialogue. Europe's Future Energy Security*. Abingdon: Routledge.

Abdelal, Rawi. 2013. The profits of power: commerce and realpolitik in Eurasia. *Review of International Political Economy* 20 (3):421–456.

AFP. 2013. Poland anchors energy strategy in coal, shale gas: PM. 10 September.

AFP. 2015. Bulgaria drops $4bn Westinghouse nuclear deal. 1 April.

Alcorn, Jessica, John Rupp and John D. Graham. 2017. Attitudes toward 'fracking': perceived and actual geographic proximity. *Review of Policy Research* 34:504–536.

Alvarez, Ramón A., Stephen W. Pacala, James J. Winebrake, William L. Chameides and Steven P. Hamburg. 2012. Greater focus needed on methane leakage from natural gas infrastructure. *Proceedings of the National Academy of Sciences* 109:6435–6440.

Andersen, Svein, Andreas Goldthau and Nick Sitter, eds. 2017. *Energy Union: Europe's New Liberal Mercantilism?* Basingstoke: Palgrave Macmillan.

Anderson, Christopher J. and Yuliya V. Tverdova. 2003. Corruption, political allegiances, and attitudes toward government in contemporary democracies. *American Journal of Political Science* 47 (1):91–109.

Andrews-Speed, Philip. 2014. The governance of unconventional gas development outside the United States of America. *Oil, Gas & Energy Law Intelligence* 12 (3).

Arnold, Gwen and Robert Holahan. 2014. The federalism of fracking: how the locus of policy-making authority affects civic engagement. *Publius: The Journal of Federalism* 44 (2):344–368.

Astrov, Alexander. 2011. *The Great Power (Mis)Management: The Russian-Georgian War and its Implications for Global Political Order*. Abingdon: Routledge.

Atkinson, Michael M. and William D. Coleman. 1989. Strong states and weak states: sectoral policy networks in advanced capitalist economies. *British Journal of Political Science* 19:47–67.

Atkinson, Michael M. and William D. Coleman. 1992. Policy networks, policy communities and the problems of governance. *Governance* 5 (2):154–180.

Avocatnet.ro. 2012. Citeste AICI textul integral al motiunii de cenzura care a dat jos Guvernul Ungureanu. Available from www.avocatnet.ro/content/articles/id_28591/Citeste-AICI-textul-integral-al-motiunii-de-cenzura-care-a-dat-jos-Guvernul-Ungureanu.html [accessed 30 October 2017].

Avritzer, Leonardo. 2009. *Participatory Institutions in Democratic Brazil*. Baltimore, MD: Johns Hopkins University Press.

Baev, Pavel K. 2008. *Russian Energy Policy and Military Power: Putin's Quest for Greatness*. London: Routledge.

Bahgat, Gawdat. 2003. Pipeline diplomacy: the geopolitics of the Caspian Sea region. *International Studies Perspectives* 3 (3):310–327.

Balmaceda, Margarita M. 2008. *Energy Dependency, Politics and Corruption in the Former Soviet Union*. New York, NY: Routledge.

Bamberger, Michelle and Robert Oswald. 2015. *The Real Cost of Fracking: How America's Shale Gas Boom Is Threatening Our Families, Pets, and Food*. Boston, MA: Beacon Press.

Baumgartner, Frank R. and Bryan D. Jones. 1991. Agenda dynamics and policy subsystems. *Journal of Politics* 53 (4):1044–1074.

Baumgartner, Frank R. and Bryan D. Jones. 1993. *Agendas and Instability in American Politics*. Chicago, IL: University of Chicago Press.

Baumgartner, Frank R. and Bryan D. Jones. 2002. Positive and negative feedback in politics. In *Policy Dynamics*, edited by F. R. Baumgartner and B. D. Jones. Chicago, IL: University of Chicago Press, pp. 3–28.

Beleva, Iskra and Mariana Kotzeva. 2001. Bulgaria: country study on international skilled migration. Available from https://mpra.ub.uni-muenchen.de/47564/ [accessed 30 October 2017].

Belyi, Andrei and Andreas Goldthau. 2015. Between a rock and a hard place: international market dynamics, domestic politics and Gazprom's strategy. EUI Working Paper RSCAS 2015/22.

Bijker, Wiebe. 1997. *Of Bicycles, Bakelites, and Bulbs: Toward a Theory of Sociotechnical Change* Cambridge, MA: MIT Press.

Bilgin, Mert. 2009. Geopolitics of European natural gas demand: Supplies from Russia, Caspian and the Middle East. *Energy Policy* 37 (11):4482–4492.

Blackwill, Robert D. and Jennifer M. Harris. 2016. *War by Other Means: Geoeconomics and Statecraft*. Cambridge, MA: Harvard University Press.

Blackwill, Robert D. and Meghan O'Sullivan. 2014. America's energy edge. The geopolitical consequences of the shale revolution. *Foreign Affairs* (March/April).

Bleich, Erik. 2002. Integrating ideas into policy-making analysis: frames and race policies in Britain and France. *Comparative Political Studies* 35 (9):1054–1076.

Bloomberg. 2012. Polish veto won't prevent EU CO2 measures, Hedegaard says. 10 March.

Bloomberg. 2014. San Leon nears first commercial European shale gas in Poland. 23 January.

BMI Research. 2015. Poland's shale progress put in danger as Chevron exits. 9 February. Available from www.bmiresearch.com/articles/polands-shale-progress-put-in-danger-as-chevron-exits [accessed 30 October 2017].

Boersma, Tim. 2015. *Energy Security and Natural Gas Markets in Europe – Lessons from the EU and the United States*. London: Routledge.

Boersma, Tim. 2016. Natural gas in the United States in 2016. Problem child and poster child. Brookings. Policy Brief 16-02.

Boersma, Tim and Andreas Goldthau. 2014. The 2014 Ukraine-Russia crisis: implications for energy markets and scholarship. *Energy Research and Social Science* 1 (3):13–15.

Boersma, Tim and Andreas Goldthau. 2017. Whither the EU's market making project in energy: from liberalization to securitization? In *Energy Union. Europe's New Liberal Mercantilism?*, edited by S. Andersen, A. Goldthau and N. Sitter. Basingstoke: Palgrave Macmillan, pp. 99–114.

Boersma, Tim and Corey Johnson. 2012. The shale gas revolution: US and EU policy and research agendas. *Review of Policy Research* 29 (4):570–576.

Boersma, Tim and Cameron Khodabakhsh. 2014. EU engagement with shale gas. *Oil, Gas and Energy Law* 3:1–11.

Bomberg, Elizabeth. 2015. Shale we drill? Discourse dynamics in UK fracking debates. *Journal of Environmental Policy & Planning* 19:1–17.

Bomberg, Elizabeth. 2017. Fracking and framing in transatlantic perspective: a comparison of shale politics in the US and European Union. *Journal of Transatlantic Studies* 15 (2):101–120.

Bordo, Michael D. and Anna Schwartz. 1997. Monetary policy regimes and economic performance: the historical record. NBER Working Paper 6201.

Boudet, Hilary Schaffer, Christopher Clarke, Dylan Bugden, Edward Maibach, Connie Roser-Renouf and Anthony Leiserowitz. 2014.

'Fracking' controversy and communication: Using national survey data to understand public perceptions of hydraulic fracturing. *Energy Policy* 65 (0):57–67.

Boutilier, Robert G. and Michal Zdziarski. 2017. Managing stakeholder networks for a social license to build. *Construction Management and Economics* 35:1–16.

Bouzarovski, Stefan. 2013. Energy poverty in the European Union: landscapes of vulnerability. *Wiley Interdisciplinary Reviews: Energy and Environment* 3 (3):276–289.

BP. 2015. BP Statistical Review of World Energy 2015. London.

BP. 2016. BP Statistical Review of World Energy. London.

BP. 2017. BP Energy Outlook. London.

Brandt, A.R., G. A. Heath, E. A. Kort, et al. 2014. Methane Leaks from North American Natural Gas Systems. *Science* 343 (6172):733–735.

Bros, Thierry. 2012. *After the US Shale Gas Revolution*. Paris: Editions Technip.

Brown, Erica, Christopher Borick, Barry G. Rabe and Thomas Ivacko. 2013. Public opinion on fracking: perspectives from Michigan and Pennsylvania. *Issues in Energy and Environmental Policy* 3 (May).

Burger, Michael. 2013. Response, fracking and federalism choice. *University of Pennsylvania Law Review* 161 (431):150–163.

Burr, Michael T. 2005. The geopolitical risks of LNG. *Public Utilities Fortnightly*.

Business Magazin. 2013. Chevron incepe cautarea gazului de sist romanesc. 28 March.

Butler, Eamonn. 2011. The geopolitics of merger and acquisition in the Central European energy market. *Geopolitics* 16 (3):626–654.

Buzogány, Aron. 2009. Romania: environmental governance – form without substance. In *Coping with Accession to the European Union: New Modes of Environmental Governance*, edited by T. A. Börzel. London: Palgrave Macmillan, pp. 169–191.

Capano, Giliberto, Michael Howlett and M. Ramesh. 2015. Bringing governments back in: governance and governing in comparative policy analysis. *Journal of Comparative Policy Analysis: Research and Practice* 17 (4):311–321.

Carmine, Joanne, Nicole Darnall and Joao Mil-Homens. 2003. Stakeholder involvement in the design of US voluntary environmental programs: does sponsorship matter? *Policy Studies Journal* 31 (4):527–543.

Castle, Ben. 2012. *The Global Movement Against Fracking. Lessons from Bulgaria, the UK and New York State*. San Francisco, CA: The Democracy Center.

Christensen, Tom and Per Lægreid. 2007. The whole-of-government approach to public sector reform. *Public Administration Review* 67 (6):1059–1066.

Christie, Edward. 2009. Energy vulnerability and EU-Russia energy relations. *Journal of Contemporary European Research* 5 (2):274–292.

Cohen, Ariel. 2009. Russia: the flawed energy superpower. In *Energy Security Challenges for the 21st Century*, edited by G. Luft and A. Korin. Westport, CT: Greenwood Publishing Group, pp. 91–108.

Colgan, Jeff. 2013. *Petro-Aggression: When Oil Causes War*. Cambridge: Cambridge University Press.

Cooke, Bill and Uma Kothari. 2001. *Participation: The New Tyranny*? New York, NY: Zed Books.

Correlje, Aad and Coby van der Linde. 2006. Energy supply security and geopolitics: a European perspective. *Energy Policy* 34 (5):532–543.

Cotton, Matthew, Imogen Rattle and James Van Alstine. 2014. Shale gas policy in the United Kingdom: an argumentative discourse analysis. *Energy Policy* 73:427–438.

Cox, Robert Henry and Daniel Béland. 2012. Valence, policy ideas, and the rise of sustainability. *Governance* 26 (2):307–328.

Crozier, Michael P. 2010. Rethinking systems: configurations of politics and policy in contemporary governance. *Administration and Society* 42:504–525.

Czyzewski, Adam B., Eduard Bodnari and Grzegorz Kozieja. 2012. *Gas (R)Evolution in Poland: Which Way to Success*? Warsaw: PKN Orlen.

Dąborowski, Tomasz and Jakub Groszkowski. 2012. Shale Gas in Bulgaria, the Czech Republic and Romania: Political Context – Legal Status – Outlook. OSW Report. Available from http://aei.pitt.edu/58010/1/shale_gas_in_bulgaria_the_czech_republic_and_romania_net_0.pdf [accessed 22 November 2017].

Dahl, Robert A. 1998. *On Democracy*. New Haven, CT: Yale University Press.

Davies, Jonathan S. 2002. Urban regime theory: a normative-empirical critique. *Journal of Urban Affairs* 24 (1):1–17.

Davies, Jonathan S. 2009. The limits of joined-up government: toward a political analysis. *Public Administration* 87 (1):80–96.

Davis, Charles. 2012. The politics of 'fracking': regulating natural gas drilling practices in Colorado and Texas. *Review of Policy Research* 29 (2):177–191.

Davis, Charles. 2014. Substate federalism and fracking policies: does state regulatory authority trump local land use autonomy? *Environmental Science & Technology* 48 (15):8397–8403.

Davis, Charles and Katherine Hoffer. 2012. Federalizing energy? Agenda change and the politics of fracking. *Policy Sciences* 45:221–241.

de Melo-Martín, Inmaculada, Jake Hays and Madelon L. Finkel. 2014. The role of ethics in shale gas policies. *Science of the Total Environment* 470:1114–1119.

Departmentul Pentru Energie. 2014. Strategia Energetica a Romaniei. Draft Supus Dezbaterii Publice. Bucharest.

Dollar, David and Victoria Levin. 2005. Sowing and Reaping: Institutional Quality and Project Outcomes in Developing Countries. World Bank Policy Research Working Paper 3524.

Dowding, Keith. 2001. Explaining urban regimes. *International Journal of Urban and Regional Research* 25 (1):7–19.

Drezner, Daniel W. 2007. *All Politics Is Global: Explaining International Regulatory Regimes*. Princeton, NJ: Princeton University Press.

Dudău, Radu. 2014. Romania's Energy Strategy Options: Current Trends in Eastern Europe's Natural Gas Markets. Bucharest: Energy Policy Group.

The Economist. 2011. Fracking heaven: other europeans fear fracking. Poland is steaming ahead. 23 June.

The Economist. 2013a. Bulgaria's electricity prices. Protesting about power prices. 15 February.

The Economist. 2013b. Shale gas in Poland. Mad and messy regulation. 10 July.

EFET. 2016. Consultation on Gas Market Development/Ankieta dotycząca oceny rynku gazu. EFET comments – 29 February 2016. Amsterdam: European Federation of Energy Traders.

EIA. 2008. Annual Energy Outlook. Washington, DC: Energy Information Agency, Department of Energy.

EIA. 2011a. Most electric generating capacity additions in the last decade were natural gas-fired. *Today in Energy*. 5 July.

EIA. 2011b. World Shale Gas Resources: An Initial Assessment of 14 Regions Outside the United States. Washington DC: Energy Information Agency, Department of Energy.

EIA. 2015a. Shale gas development in China aided by government investment and decreasing well cost. *Today in Energy*. 30 September.

EIA. 2015b. Technically Recoverable Shale Oil and Shale Gas Resources: Other Eastern Europe, Washington, DC: Energy Information Agency, Department of Energy.

EIA. 2016a. Energy-related CO2 emissions for first six months of 2016 are lowest since 1991. *Today in Energy*.12 October.

EIA. 2016b. US Shale Production. US Department of Energy. Available from https://www.eia.gov/dnav/ng/hist/res_epg0_r5302_nus_bcfa.htm [accessed 5 May 2017].

EIA. 2017a. Annual Energy Outlook 2017. Washington, DC: US Department of Energy.
EIA. 2017b. Argentina seeking increased natural gas production from shale resources to reduce imports. *Today in Energy*. 17 February.
EIA. 2017c. Electricity Monthly Update. Washington, DC: Energy Information Agency, Department of Energy.
EIA. 2017d. Natural Gas Gross Withdrawals and Production. Available from https://www.eia.gov/dnav/ng/ng_prod_sum_a_EPG0_FPD_mmcf_a.htm [accessed 18 July 2017].
EIA. 2017e. Short Term Energy Outlook (STEO). Washington, DC: Energy Information Agency, Department of Energy.
EIA. 2017f. US Dry Natural Gas Production. Available from https://www.eia.gov/dnav/ng/hist/n9070us2m.htm [accessed 22 May 2017].
EIA/ARI. 2013. EIA/ARI World Shale Gas and Shale Oil Resource Assessment. Technically Recoverable Shale Oil and Shale Gas Resources: An Assessment of 137 Shale Formations in 41 Countries Outside the United States. Washington, DC: Department of Energy.
Eisner, Marc Allen. 1994. Discovering patterns in regulatory history: continuity, change and regulatory regimes. *Journal of Policy History* 6 (2):157–187.
Eisner, Marc Allen. 2000. *Regulatory Politics in Transition*, 2nd edn. Baltimore, MD: Johns Hopkins University Press.
Elkin, Stephen L. 1987. *City and Regime in the American Republic*. Chicago, IL: University of Chicago Press.
Elmore, Richard F. 1978. Organizational models of social program implementation. *Public Policy* 26 (2):185–226.
EMIS. 2014. Energy Sector Poland. London/New York/Mumbai.
Energy Charter Secretariat. 2007. Putting a Price on Energy. International Pricing Mechanisms for Oil and Gas. Brussels.
Energy and Water Regulatory Commission. 2010. 2009 National Report to the European Commission. Sofia.
Energyworldmag. 2016. Romania: ExxonMobil, OMV Petrom Complete Black Sea Drilling. 10.02.
Engeli, Isabelle and Christine Rothmayr Allison, eds. 2014. *Comparative Policy Studies: Conceptual And Methodological Challenges*. Basingstoke: Palgrave Macmillan.
ENTSO-G. 2015. Ten Year Network Development Plan 2015. Brussels.
EPG. 2014. The Need to Update the Romanian Oil and Gas Legislation for Exploration and Production, Bucharest: Energy Policy Group.
Ernst & Young. 2013. Analysis of Central Europe's Energy Sector. Warsaw.
Esping-Anderson, Gosta. 1990. *The Three Worlds of Welfare Capitalism*. Princeton, NJ: Princeton University Press.

EUObserver. 2012. EU ditches plan to regulate on shale gas. 22 January.

Euractiv. 2012a. Hungary to step up nationalisations. 5 September.

Euractiv. 2012b. Poland defies Europe over 2050 low-carbon roadmap. 8 March.

Euractiv. 2012c. US tells Bulgaria shale gas is safe. 7 February.

Euractiv. 2013. In historic vote, Bulgarian voters back new nuclear plant. 28 January.

Euractiv. 2014. Poland says it 'won' at the EU summit. 24 October.

Euractiv. 2015. Bulgaria lacks political will to build interconnectors, says Commission. 6 March.

Eurocoal. 2017. Country profiles. Available from https://euracoal.eu/info/country-profiles/poland/ [accessed 17 June 2017].

Eurogas. 2013. Long-Term Outlook for Gas to 2035. Brussels.

Eurogas. 2014. Statistical Report 2014. Brussels.

European Commission. 2003. Commission Reaches Breakthrough with Gazprom and ENI on Territorial Restriction Clauses (IP/03/1345). Brussels.

European Commission. 2009. Directive 2009/28/EC of the European Parliament and of the Council of 23 April 2009 on the promotion of the use of energy from renewable sources and amending and subsequently repealing Directives 2001/77/EC and 2003/30/EC (Text with EEA relevance). *Official Journal of the European Communities L* 140 (5 June):16–61.

European Commission. 2010. Energy 2020. A Strategy for Competitive, Sustainable and Secure Energy. Brussels: COM(2010) 639 COM(2010) 639.

European Commission. 2011a. Energy Roadmap 2050. Brussels: COM(2011) 885 final.

European Commission. 2011b. EU Gas Market: Commission Refers Bulgaria and Romania to Court to Ensure European Law is Properly Implemented IP/11/1437. 24 November.

European Commission. 2012a. Press Release. Antitrust: Commission Opens Proceedings Against Gazprom. 4 September.

European Commission. 2012b. Quarterly Report: Energy on European Gas Markets. Fourth Quarter 2012. Brussels.

European Commission. 2013a. Antitrust: Commission Opens Proceedings Against Bulgarian Energy Holding and its Subsidiaries Bulgargaz and Bulgartransgaz. 5 July.

European Commission. 2013b. Enforcing EU Energy Law Gas & Electricity. Infringement Proceedings. Available from http://ec.europa.eu/energy/infringements/proceedings/electricity_gas_en.htm [accessed 30 May 2013].

European Commission. 2013c. Flash Eurobarometer 360: Attittudes of Europeans Towards Air Quality. Brussels.
European Commission. 2013d. Internal Energy Market: Commission Refers Bulgaria, Estonia and the United Kingdom to Court for Failing to Fully Transpose EU Rules IP/13/42. 24 January.
European Commission. 2013e. Report to the Bulgarian Government on 'Findings and Recommendations Related to Bulgarian Energy Policy'. Sofia.
European Commission. 2014a. Commission Recommendation of 22 January 2014 on minimum principles for the exploration and production of hydrocarbons (such as shale gas) using high-volume hydraulic fracturing. *Official Journal L* 39 (8 February):72–78.
European Commission. 2014b. Commission Recommendation of 22 January 2014 on minimum principles for the exploration and production of hydrocarbons (such as shale gas) using high-volume hydraulic fracturing 2014/70/EU. *Official Journal L* 39:72–78.
European Commission. 2014c. Communication from the Commission to the Council and the European Parliament on the Exploration and Production of Hydrocarbons (such as Shale Gas) Using High Volume Hydraulic Fracturing in the EU. Brussels: COM (2014) 23.
European Commission. 2014d. Communication from the Commission to the European Parliament and the Council on the Short Term Resilience of the European Gas System. Preparedness for a Possible Disruption of Supplies from the East during the Fall and Winter of 2014/2015. Brussels: COM(2014) 654 final.
European Commission. 2014e. Communication from the Commission to the European Parliament, the Council, the European Economic and Social Committee and the Committee of the Regions. A Policy Framework for Climate and Energy in the Period from 2020 to 2030. Brussels: COM/2014/015 final.
European Commission. 2014f. Country Report: Poland. Brussels.
European Commission. 2014g. Country Report: Bulgaria. Brussels.
European Commission. 2014h. Country Report: Romania. Brussels.
European Commission. 2014i. EU Energy, Transport and GHG Emissions Trends to 2050. Brussels: DG Energy.
European Commission. 2014j. Macroeconomic Impacts of Shale Gas Extraction in the EU. Brussels: DG ENV.ENV.F.1/SER/2012/0046r.
European Commission. 2014k. Quarterly report on European gas markets. *Market Observatory for Energy* 9(7).
European Commission. 2015a. Agriculture and Rural Development. Member States Factsheets: Bulgaria. Brussels.

European Commission. 2015b. Report from the Commission to the European Parliament and the Council. Assessment of the Progress Made by Member States Towards the National Energy Efficiency Targets for 2020 and Towards the Implementation of the Energy Efficiency Directive 2012/27/EU as Required by Article 24 (3) of Energy Efficiency Directive 2012/27/EU. Brussels: COM(2015) 574 final.

European Commission. 2016a. Antitrust: Commission Confirms Unannounced Inspections in the Natural Gas Sector. Brussels: Statement/16/2133. 7 June.

European Commission. 2016b. Communication from the Commission to the European Parliament, the COuncil, the European Economic and Social Comittee, the Committee of the Regions and the European Investment Bank. Clean Energy for All Europeans. Brussels: COM(2016) 860 final.

European Commission. 2016c. Country Report Romania 2016 Including an In-Depth Review on the Prevention and Correction of Macroeconomic Imbalances. Brussels: Commission Staff Working Document SWD(2016) 91 final.

European Commission. 2016d. Projects of Common Interest. Available from https://ec.europa.eu/energy/en/topics/infrastructure/projects-common-interest [accessed 12 April 2016].

European Commission. 2016e. Proposal for a Regulation of the European Parliament and of the Council on the Governance of the Energy Union, Amending Directive 94/22/EC, Directive 98/70/EC, Directive 2009/31/EC, Regulation (EC) No 663/2009, Regulation (EC) No 715/2009, Directive 2009/73/EC, Council Directive 2009/119/EC, Directive 2010/31/EU, Directive 2012/27/EU, Directive 2013/30/EU and Council Directive (EU) 2015/652 and Repealing Regulation (EU) No 525/2013 (Text with EEA Relevance). Brussels: COM(2016) 759 final.

European Commission. 2016f. Quarterly Report. Energy on European Gas Markets. Fourth Quarter of 2015 and First Quarter of 2016. Brussels.

European Commission. 2017. New Governance to Deliver on Objectives of the Energy Union. Available from https://ec.europa.eu/energy/en/new-governance-deliver-objectives-energy-union [accessed 17 June 2017].

European Communities. 2007. The Treaty of Lisbon Amending the Treaty Establishing the European Union and the Treaty Establishing the European Community Including the Protocols and Annexes, and Final Act with Declarations. Brussels: C 306 (01).

European Council. 1992a. Council Directive 92/43/EEC of 21 May 1992 on the conservation of natural habitats and of wild fauna and flora. *Official Journal* L 206:7–50.

European Council. 1992b. Council Directive 92/91/EEC of 3 November 1992 concerning the minimum requirements for improving the safety and health protection of workers in the mineral- extracting industries through drilling (eleventh individual Directive within the meaning of Article 16 (1) of Directive 89/391/EEC). *Official Journal L* 348:9–24.

European Court of Justice. 2013. Judgment of the Court (Fourth Chamber) of 27 June 2013. *European Commission* v. *Republic of Poland*. Failure of a Member State to Fulfil Obligations – Directive 94/22/EC – Conditions for Granting and Using Authorisations for the Prospection, Exploration and Extraction of Hydrocarbons – Non-Discriminatory Access. Luxembourg: Case C-569/10.

European Parliament and the Council. 1994. Directive 94/22/EC of the European Parliament and of the Council of 30 May 1994 on the conditions for granting and using authorizations for the prospection, exploration and production of hydrocarbons. *Official Journal L* 164:3–8.

European Parliament and the Council. 1998. Directive 98/30/EC Concerning Common Rules for the Internal Market in Natural Gas. Brussels: European Parliament and Council of the European Union, 22 June.

European Parliament and the Council. 2000. Directive 2000/60/EC of the European Parliament and of the Council establishing a framework for the Community action in the field of water policy. *Official Journal L* 327:1–73.

European Parliament and the Council. 2001. Directive 2001/42/EC of the European Parliament and of the Council of 27 June 2001 on the assessment of the effects of certain plans and programmes on the environment. *Official Journal L* 197:30–37.

European Parliament and the Council. 2003. Directive 2003/55/EC Concerning Common Rules for the Internal Market in Natural Gas and Repealing Directive 98/30/EC. Brussels: European Parliament and Council of the European Union, 26 June.

European Parliament and the Council. 2004. Directive 2004/35/CE of the European Parliament and of the Council of 21 April 2004 on environmental liability with regard to the prevention and remedying of environmental damage. *Official Journal L* 143:56–75.

European Parliament and the Council. 2006a. Directive 2006/21/EC of the European Parliament and of the Council on the management of waste from the extractive industries. *Official Journal L* 102:1–29.

European Parliament and the Council. 2006b. Directive 2006/118/EC of the European Parliament and of the Council of 12 December 2006 on the protection of groundwater against pollution and deterioration. *Official Journal L* 372:19–31.

European Parliament and the Council. 2006c. Directive 2006/121/EC of the European Parliament and of the Council amending Council Directive 67/548/EEC on the approximation of laws, regulations and administrative provisions relating to the classification, packaging and labelling of dangerous substances in order to adapt it to Regulation (EC) No. 1907/2006 concerning the Registration, Evaluation, Authorisation and Restriction of Chemicals (REACH) and establishing a European Chemicals Agency. *Official Journal L* 396:850–856.

European Parliament and the Council. 2006d. Regulation (EC) No 1907/2006 of the European Parliament and of the Council of 18 December 2006 concerning the Registration, Evaluation, Authorisation and Restriction of Chemicals (REACH), establishing a European Chemicals Agency, amending Directive 1999/45/EC and repealing Council Regulation (EEC) No 793/93 and Commission Regulation (EC) No 1488/94 as well as Council Directive 76/769/EEC and Commission Directives 91/155/EEC, 93/67/EEC, 93/105/EC and 2000/21/EC. *Official Journal L* 49:396.

European Parliament and the Council. 2009a. Decision No 406/2009/EC of the European Parliament and of the Council of 23 April 2009 on the effort of Member States to reduce their greenhouse gas emissions to meet the Community's greenhouse gas emission reduction commitments up to 2020. *Official Journal L* 140:136–148.

European Parliament and the Council. 2009b. Directive 2009/29/EC of the European Parliament and of the Council of 23 April 2009 amending Directive 2003/87/EC so as to improve and extend the greenhouse gas emission allowance trading scheme of the Community. *Official Journal L* 140:63–87.

European Parliament and the Council. 2009c. Directive 2009/31/EC of the European Parliament and of the Council of 23 April 2009 on the geological storage of carbon dioxide and amending Council Directive 85/337/EEC, European Parliament and Council Directives 2000/60/EC, 2001/80/EC, 2004/35/EC, 2006/12/EC, 2008/1/EC and Regulation (EC) No 1013/2006 (Text with EEA relevance). *Official Journal L* 140:114–134.

European Parliament and the Council. 2009d. Directive 2009/73/EC of the European Parliament and of the Council of 13 July 2009 Concerning Common Rules for the Internal Market in Natural Gas and Repealing Directive 2003/55/EC. Brussels: European Parliament and Council of the European Union.

European Parliament and the Council. 2010a. Directive 2009/147/EC of the European Parliament and of the Council of 30 November 2009 on the conservation of wild birds. *Official Journal L* 20:7–25.

European Parliament and the Council. 2010b. Directive 2010/75/EU of the European Parliament and of the Council of 24 November 2010 on industrial emissions (integrated pollution prevention and control). Text with EEA relevance. *Official Journal L* 334:17–119.

European Parliament and the Council. 2010c. Regulation (EU) No 994/2010 of the European Parliament and of the Council of 20 October 2010 concerning measures to safeguard security of gas supply and repealing Council Directive 2004/67/EC (Text with EEA relevance). *Official Journal of the European Union L* 295:1–22.

European Parliament and the Council. 2012a. Directive 2011/92/EU of the European Parliament and of the Council of 13 December 2011 on the assessment of the effects of certain public and private projects on the environment (codification). Text with EEA relevance. *Official Journal L* 26:1–21.

European Parliament and the Council. 2012b. Directive 2012/18/EU of the European Parliament and of the Council of 4 July 2012 on the control of major-accident hazards involving dangerous substances, amending and subsequently repealing Council Directive 96/82/EC. Text with EEA relevance. *Official Journal L* 197:1–37.

European Parliament and the Council. 2013. Directive 2013/30/EU of the European Parliament and of the Council of 12 June 2013 on safety of offshore oil and gas operations and amending Directive 2004/35/EC. Text with EEA relevance. *Official Journal L* 178:66–106.

European Parliament and the Council. 2014. Directive 2014/52/EU of the European Parliament and the Council 16 April 2014 amending Directive 2011/92/EU on the assessment of the effects of certain public and private projects on the environment (Text with EEA relevance). *Official Journal L* 124:1–18.

Eurostat. 2014. Renewable Energy Statistics. Brussels.

Eurostat. 2015. GDP Per Capita at Current Market Prices, 2005 and 2015. Available from http://ec.europa.eu/eurostat/statistics-explained/index.php/File:GDP_per_capita_at_current_market_prices,_2005_and_2015_(EU-28_%3D_100;_based_on_PPS_per_inhabitant)_YB16.png [accessed 11 November 2016].

Eurostat. 2016. Energy Production and Imports. July. Available from http://ec.europa.eu/eurostat/statistics-explained/index.php/Energy_production_and_imports [accessed 2 May 2017].

Eurostat. 2017. EU Imports of Energy Products – Recent Developments. Available from http://ec.europa.eu/eurostat/statistics-explained/index.php/EU_imports_of_energy_products_-_recent_developments [accessed 1 June 2017].

Evensen, Darrick. 2016. Ethics and 'fracking': a review of (the limited) moral thought on shale gas development. *Wiley Interdisciplinary Reviews: Water* 3 (4):575–586.

Evensen, Darrick T., Christopher E. Clarke and Richard C. Stedman. 2014. A New York or Pennsylvania state of mind: social representations of gas development in the Marcellus Shale. *Journal of Environmental Studies and Sciences* 4:65–77.

Fagan, Adam and JoAnn Carmin, eds. 2011. *Green Activism in Post-Socialist Europe and the Former Soviet Union*. London: Routledge.

Financial Times. 2010. Europe the new frontier in shale gas rush. 7 March.

Financial Times. 2014a. Bulgarians see Russian hand in anti-shale protests. 30 November.

Financial Times. 2014b. Nato claims Moscow funding anti-fracking groups. 19 June.

Financial Times. 2014c. ENI joins shale gas exodus from Poland. 15 January.

Financial Times. 2015a. Eastern European shale exploration on ice as boom turns to bust. 28 October.

Financial Times. 2015b. The US shale revolution. 24 April.

Finon, Dominique and Catherine Locatelli. 2008. Russian and European gas interdependence. Can market forces balance out geopolitics? *Energy Policy* 36 (1):423–442.

Freeman, J. Leiper and Judith Parris Stevens. 1987. A theoretical and conceptual reexamination of subsystem politics. *Public Policy and Administration* 2 (9):9–24.

Gabor, Hunya. 2007. Privatization Disputes in Romania – the Petrom Case, Vienna: Vienna Institute for International Economic Studies. wiiw Research Report Nr. 337.

Garpiel, Rafał. 2014. Attitudes towards shale gas in Poland and the Czech Republic. In *The Development of the Shale Gas Sector in Poland and its Prospects in the Czech Republic – Analysis and Recommendations*, edited by I. Albrycht, W. Bigaj, V. Dvořáková, et al. Krakow: The Kosciuszko Institute.

Gazprom Export. 2016. Gas Supplies to Europe. Available from www.gazpromexport.ru/en/statistics/ [accessed 19 May 2016].

Geels, Frank W. 2005. Processes and patterns in transitions and system innovations: Refining the co-evolutionary multi-level perspective. *Technological Forecasting & Social Change* 72:681–696.

Geels, Frank W. and René Kemp. 2007. Dynamics in socio-technical systems: typology of change processes and contrasting case studies. *Technology and Society* 29 (4):441–455.

Gény, Florence. 2010. Can Unconventional Gas be a Game Changer in European Gas Markets? Oxford Institute for Energy Studies. NG 46.

Georgiev, Atanas. 2011. Shale gas battle in Bulgaria – high stalkes for Europe. *European Energy Review*. 1 September.

Georgiev, Atanas. 2016. Shale and Eastern Europe – Bulgaria, Romania, and Ukraine. In *The Global Impact of Unconventional Shale Gas Development*, edited by Y. Wang and W. E. Hefley. New York, NY: Springer, 75–96.

Georgieva, Patricia. 2004. Bulgaria: the double edge of economy and demography. *Higher Education in Europe* 29 (3):363–372.

Giurco, Damien, Benjamin Mclellan, Daniel M. Franks, Keisuke Nansai and Timothy Prior. 2014. Responsible mineral and energy futures: views at the nexus. *Journal of Cleaner Production* 84:322–338.

Godzimirski, Jakub M. and Krzysztof Kasianiuk, eds. 2014. *Polish and Norwegian Governance: Closing the Gaps*. Warsaw: Polski Instytut Spraw Międzynarodowych (PISM).

Gold, David A., Clarence Y. H. Lo and Erik Olin Wright. 1975. Recent developments in Marxist theories of the capitalist state. *Monthly Review* 27 (5):29–43.

Gold, Russell. 2014. *The Boom: How Fracking Ignited the American Energy Revolution and Changed the World*. New York, NY: Simon & Schuster.

Goldthau, Andreas. 2008. Resurgent Russia? Rethinking Energy Inc. Five myths about the 'energy superpower'. *Policy Review* 147 (February/March):53–63.

Goldthau, Andreas. 2011a. Challenges in global oil governance. In *Oil Politics Handbook*, edited by R. Looney. London: Routledge, pp. 348–360.

Goldthau, Andreas. 2011b. A public policy perspective on global energy security. *International Studies Perspectives* 13 (December): 64–83.

Goldthau, Andreas. 2012a. Emerging governance challenges for Eurasian gas markets after the shale gas revolution. In *Dynamics of Energy Governance in Europe and Russia*, edited by C. Kuzemko, A. Belyi, A. Goldthau and M. Keating. Basingstoke: Palgrave Macmillan.

Goldthau, Andreas. 2012b. From the state to the market and back. Policy implications of changing energy paradigms. *Global Policy* 2 (S2):198–210.

Goldthau, Andreas. 2013. *The politics of natural gas development in the European Union*. Cambridge, MA/Houston, TX: Harvard Belfer Center/Rice University's Baker Institute.

Goldthau, Andreas. 2016a. Assessing Nord Stream 2: Regulation, Geopolitics & Energy Security in the EU, Central Eastern Europe & the UK. London: King's College London, Russia Institute & EUCERS. EUCERS Strategy Paper 10.

Goldthau, Andreas. 2016b. Conceptualizing the above ground factors in shale gas. Toward a research agenda on regulatory governance. *Energy Research and Social Science* 20:73–81.

Goldthau, Andreas. 2016c. Energy technology, politics, and interpretative frames: the case of shale gas fracking in Eastern Europe. *Global Environmental Politics* 16 (4):50–69.

Goldthau, Andreas. 2017. The G20 must govern the shift to low-carbon energy. *Nature* 546:203–205.

Goldthau, Andreas and Michael LaBelle. 2016. The power of policy regimes. Explaining shale gas policy divergence in Bulgaria and Poland. *Review of Policy Research* 33 (6):603–622.

Goldthau, Andreas and Nick Sitter. 2014. A liberal actor in a realist world? The Commission and the external dimension of the single market for energy. *Journal of European Public Policy* 21 (10):1452–1472

Goldthau, Andreas and Nick Sitter. 2015a. *A Liberal Actor in a Realist World. The European Union Regulatory State and the Global Political Economy of Energy*. Oxford: Oxford University Press.

Goldthau, Andreas and Nick Sitter. 2015b. Soft power with a hard edge: EU policy tools and energy security. *Review of International Political Economy* 22 (5):941–965.

Goldthau, Andreas and Benjamin Sovacool. 2016. Energy technology, politics, and interpretative frames: The case of shale gas fracking in Eastern Europe. *Global Environmental Politics* 16 (4):50–69.

Goldthau, Andreas and Jan Martin Witte. 2009. Back to the future or forward to the past? Strengthening markets and rules for effective global energy governance. *International Affairs* 85 (2):373–390.

Grafton, R. Quentin, Ian G. Cronshaw and Michal C. Moore, eds. 2017. *Risks, Rewards and Regulation of Unconventional Gas: A Global Perspective*. Cambridge: Cambride University Press.

Graham, John D., John A. Rupp and Olga Schenk. 2015. Unconventional Gas Development in the USA: Exploring the Risk Perception Issues. *Risk Analysis* 35 (10):1770–1788.

Graves, John H. 2012. *Fracking: America's Alternative Energy Revolution*. Asheville: Safe Harbor.

Groen, Lisanne, Arne Niemann and Sebastian Oberthür. 2012. The EU as a global leader? The Copenhagen and Cancun UN climate change negotiations. *Journal of Contemporary European Research* 8 (2):173–191.

Grubler, Arnulf and Charlie Wilson, eds. 2013. *Energy Technology Innovation: Learning from Historical Successes and Failures*. Cambridge: Cambridge University Press.

The Guardian. 2010. Ukraine extends lease for Russia's Black Sea fleet. 21 April.

The Guardian. 2012. Bulgaria becomes second state to impose ban on shale-gas exploration. 14 February.

The Guardian. 2013a. Chevron suspends shale gas exploration plan in Romanian village after protest. 21 October.

The Guardian. 2013b. Police remove protesters from Chevron's fracking site in Romania. 5 December.

The Guardian. 2013c. Romania expected to reject gold mine following week of protest. 10 September.

The Guardian. 2015. Poland's shale gas revolution evaporates in face of environmental protests. 12 January.

The Guardian. 2016a. Leaked TTIP energy proposal could 'sabotage' EU climate policy. 11 July.

The Guardian. 2016b. Romanian village blocks Canadian firm from mining for gold. 14 January.

Gullion, Jessica Smartt. 2015. *Fracking the Neighborhood. Reluctant Activists and Natural Gas Drilling*. Cambridge, MA: MIT Press.

Gunningham, Neil, Robert A. Kagan and Dorothy Thornton. 2004. Social licence and environmental protection: why businesses go beyond compliance. *Law & Social Inquiry* 29:307–341.

Gupta, Kuhika. 2012. Comparative public policy: using the comparative method to advance our understanding of the policy process. *Policy Studies Journal* 40:11–26.

Gusilov, Eugenia. 2012. Romania's Shale Gas Strategy. Bucharest: Romania Energy Center. Policy Brief 1.

Gustafson, Thane. 1998. *Crisis Amid Plenty: The Politics of Soviet Energy under Brezhnev and Gorbachev*. Princeton, NJ: Princeton University Press.

Hall, Peter A. and Rosemary C. R. Taylor. 1996. Political science and the three new institutionalisms. *Political Studies* 44 (December):936–957.

Hardt, John and Donna L. Gold. 1982. Soviet Gas Pipeline: US Options. Washington, DC: Congressional Research Service. IB82020.

Harris, Richard A. and Sidney M. Milkis. 1996. *The Politics of Regulatory Change: A Tale of Two Agencies*. New York, NY: Oxford University Press.

Haug, Marianne. 2012. Shale gas and renewables: divergence or win-win for transatlantic energy cooperation? *Journal of Transatlantic Studies* 10 (4):358–373.

Hausman, Catherine and Ryan Kellogg. 2015. *Welfare and Distributional Implications of Shale Gas*. Washington, DC: Brookings.

Head, Brian W. 2008. Wicked problems in public policy. *Public Policy* 3 (2):101–118.

Hickey, Samuel and Giles Mohan, eds. 2004. *Participation: From Tyranny to Transformation?* New York, NY: Zed Books.

Hiteva, Ralitsa Petrova and Tomas Maltby. 2014. Standing in the way by standing in the middle: the case of state-owned natural gas intermediaries in Bulgaria. *Geoforum* 54:120–131.

Högselius, Per. 2012. *Red Gas. Russia and the Origins of European Energy Dependence*. Basingstoke: Palgrave Macmillan.

Hood, Christopher, Henry Rothstein and Robert Baldwin. 2001. *The Government of Risk: Understanding Risk Regulation Regimes*. Oxford: Oxford University Press.

Horsford, Sonya Douglass, ed. 2010. *New Perspectives in Educational Leadership. Exploring Social, Political, and Community Contexts and Meaning*. New York, NY: Peter Lang.

House, Evan J. 2013. Fractured fairytales: the failed social license for unconventional oil and gas development. *Wyoming Law Review* 13 (1): 6–67.

House of Lords. 2013. The Economic Impact on UK Energy Policy of Shale Gas and Oil. Oral and Written Evidence. London. Available from https://publications.parliament.uk/pa/ld201314/ldselect/ldeconaf/172/172.pdf [accessed 22 November 2017].

Howarth, Robert W. 2015. Methane emissions and climatic warming risk from hydraulic fracturing and shale gas development: implications for policy. *Energy and Emission Control Technologies* 3:45–54.

Howarth, Robert W., Renee Santoro and Anthony Ingraffea. 2011. Methane and the greenhouse-gas footprint of natural gas from shale formations. *Climatic Change* 106 (4):679–690.

Howlett, Michael. 2001. *Canadian Forest Policy: Adapting to Change*. Toronto, ON: University of Toronto Press.

Howlett, Michael. 2004. Beyond good and evil in policy implementation: instrument mixes, implementation styles, and second generation theories of policy instrument choice. *Policy and Society* 23 (2):1–17.

Howlett, Michael. 2009. Governance modes, policy regimes, and operational plans: a multi-level nested model of policy instrument choice an policy design. *Policy Sciences* 42 (1):73–89.

Howlett, Michael and Ishani Mukherjee. 2014. Policy design and non-design: towards a spectrum of policy formulation types. *Politics and Governance* 2 (2):57–71.

Howlett, Michael and M. Ramesh. 2002. The policy effects of internationalization: a subsystem adjustment analysis of policy change *Journal of Comparative Policy Analysis* 4 (1):31–50.

Howlett, Michael and Jeremy Rayner. 2006. Convergence and divergence in 'new governance' arrangements: evidence from european integrated natural resource strategies. *Journal of Public Policy* 26 (2):167–189.

Howlett, Michael, M. Ramesh and Anthony Perl. 2009. *Studying Public Policy. Policy Cycles and Policy Subsystems*, 3rd edn. Oxford: Oxford University Press.

Hoy, Kyle A., Timothy W. Kelsey and Martin Shields. 2017. An economic impact report of shale gas extraction in Pennsylvania with stricter assumptions. *Ecological Economics* 138:178–185.

Hulbert, Matthew and Andreas Goldthau. 2013. Natural gas going global? Potential and pitfalls. In *Handbook of Global Energy Policy*, edited by A. Goldthau. London: Wiley-Blackwell, pp. 98–112.

Hultman, Nathan, Dylan Rebois, Michael Scholten and Christopher Ramig. 2011. The greenhouse impact of unconventional gas for electricity generation. *Environmental Research Letters* 6:044008.

Hunter, Janet R and Zachary A Smith. 2006. *Protecting Our Environment: Lessons from the European Union*. New York, NY: SUNY Press.

ICIS. 2011. Bulgaria planning to cut Russian natural gas imports by 75%. 7 July. Available from www.icis.com/heren/articles/2011/07/07/9475963/bulgaria+planning+to+cut+russian+natural+gas+imports+by.html [accessed 18 December 2013].

IEA. 2012a. Golden Rules for a Golden Age of Gas. World Energy Outlook Special Report on Unconventional Gas. Paris: OECD.

IEA. 2012b. Natural Gas Information. Paris: OECD.

IEA. 2014a. Energy Supply Security: The Emergency Response of IEA Countries – 2014 Edition. Paris: OECD.

IEA. 2014b. Medium-Term Gas Market Report. Paris: OECD.

IEA. 2015. World Energy Outlook. Paris: OECD.

IEA. 2016. World Energy Outlook. Paris: OECD.

IGU. 2016. Wholesale Gas Price Survey 2016 Edition. Fornebu.

IHS Energy. 2016. Shale Gas Reloaded: The Evolving View of North American Natural Gas Resources and Costs. London.

Imbroscio, David L. 1998. Reformulating urban regime theory: the division of labor between state and market reconsidered. *Journal of Urban Affairs* 20 (3):233–248.

IMF. 2014. 25 Years of Transition. Post-Communist Europe and the IMF. Washington, DC.

Inmam, Mason. 2016. Can fracking power Europe? *Nature* 532 (7592):22–24.

International Business Times. 2010. Poland: the next fracking frontier? 20 October.

International Herald Tribune. 2008. Russia, Iran and Qatar move toward forming gas cartel. 22 October.

Jackson, Robert B., Avner Vengosh, J. William Carey, et al. 2014. The environmental costs and benefits of fracking. *Annual Review of Environment and Resources* 39 (7):1–7.

Jacquet, Jeffrey B. 2014. Review of risks to communities from shale energy development. *Environmental Science & Technology* 48 (15):8321–8333.

Jaffe, Amy Myers. 2010. Shale gas will rock the world. *Wall Street Journal*. 10 May.

Jahn, Detlef. 1998. Environmental performance and policy regimes: Explaining variations in 18 OECD-countries. *Policy Sciences* 31:107–131.

Jans, Jan H. and Hans H. Vedder. 2011. *European Environmental Law: After Lisbon Vol. 4*. Groningen: Europa Law Publishing.

Jaspal, Rusi and Brigitte Nerlich. 2014. Fracking in the UK press: threat dynamics in an unfolding debate. *Public Understanding of Science* 23 (3):348–363.

Jaspal, Rusi, Brigitte Nerlich and Szczepan Lemańcyzk. 2014. Fracking in the Polish press: geopolitics and national identity. *Energy Policy* 74:253–261.

Jaspal, Rusi, Andrew Turner and Brigitte Nerlich. 2014. Fracking on YouTube: exploring risks, benefits and human values. *Environmental Values* 23 (5):501.

Jenkins-Smith, Hank C., Gilbert St Clair and Brian Woods. 1991. Explaining change in policy subsystems: analysis of coalition stability and defection over time. *American Journal of Political Science* 35 (4):851–881.

Jochim, Ashley E. and Peter J. May. 2010. Beyond subsystems: policy regimes and governance. *Policy Studies Journal* 38 (2):303–327.

Johnson, Corey and Matthew Derrick. 2012. A splintered heartland: russia, europe, and the geopolitics of networked energy infrastructure. *Geopolitics* 17 (3):482–501.

Jong, Sijbren de, Willem Auping and Joris Govers. 2014. *The Geopolitics of Shale Gas. The Implications of the US Shale Gas Revolution on Intrastate Stability within Traditional Oil- and Natural Gas Exporting Countries in the EU Neighborhood*. The Hague: The Hague Center for Strategic Studies.

Jordan, Andy and Elah Matt. 2014. Designing policies that intentionally stick: policy feedback in a changing climate. *Policy Sciences* 47 (3):227–247.

Jordan, Grant. 1990. Sub-governments, policy communities and networks: refilling the old bottles? *Journal of Theoretical Politics* 2 (3):319–338.

Jordana, Jacint, David Levi-Faur and Imma Puig. 2006. The limits of Europeanization: Regulatory reforms in the Spanish and Portuguese telecommunications and electricity sectors. *Governance* 19 (3):437–464.

Judge, Andrew and Tomas Maltby. 2017. European Energy Union? Caught between securitisation and 'riskification'. *European Journal of International Security* 2 (2):179–202.

Kalicki, Jan H. and David L. Goldwyn, eds. 2005. *Energy and Security. Toward a New Foreign Policy Strategy*. Princeton, NJ: Woodrow Wilson Center Press.

Kaplan, Robert D. 2012. The geopolitics of shale. *Stratfor Worldview*. Available from https://worldview.stratfor.com/article/geopolitics-shale [accessed 22 November 2017].

Kavanagh, D. and D. Richards. 2001. Departmentalism and joined-up government. *Parliamentary Affairs* 54 (1):1–18.

Kazepov, Yuri, ed. 2004. *Cities of Europe: Changing Contexts, Local Arrangement and the Challenge to Urban Cohesion*. London: Wiley-Blackwell.

KC2. 2014. Natural Gas from Shale. Socioeconomic Impacts for Bulgaria. Sofia: Institute for Market Economics.

Kemp, René, Jan Rotmans and Derk Loorbach. 2007. Assessing the Dutch energy transition policy: how does it deal with dilemmas of managing transitions? *Journal of Environmental Policy & Planning* 9 (3–4):315–331.

Kenarov, Dimiter. 2011. Where your gold comes from: the story of an exploited town in Bulgaria. *The Atlantic*. 7 November.

Kenarov, Dimiter. 2012. Poland's shale gas dream. *Foreign Policy*. 26 December.

Keohane, Robert O. 1984. *After Hegemony. Cooperation and Discord in the World Political Economy*. Princeton, NJ: Princeton University Press.

Kettl, Donald F. 2000. The transformation of governance: globalization, devolution, and the role of government. *Public Administration Review* 60 (6):488–497.

Kim, Younkyoo and Stephen Blank. 2014. US shale revolution and Russia: shifting geopolitics of energy in Europe and Asia. *Asia Europe Journal* 13 (1):95–112.

Kingdon, John W. 1984. *Agendas, Alternatives, and Public Policies*. Boston, MA: Little Brown.

Klare, Michael. 2001. *Resource Wars: The New Landscape of Global Conflict*. New York, NY: Henry Holt.

Klare, Michael T. 2009. *Rising Powers, Shrinking Planet: The New Geopolitics of Energy*. New York, NY: Metropolitan Books.

Klein, Hans E. and Daniel Lee Kleinman. 2002. The social construction of technology: structural considerations. *Science, Technology, and Human Values* 27 (1):28–52.

Knill, Christoph. 1998. European policies: the impact of national administrative traditions. *Journal of Public Policy* 18 (1):1–28.

Knill, Christoph and Dirk Lehmkuhl. 2002. The national impact of European Union regulatory policy: three Europeanization mechanisms. *European Journal of Political Research* 41 (2):255–280.

Kohl, Wilfrid. 2010. The International Energy Agency and the Global Energy Order. In *Global Energy Governance: The New Rules of the Game*, edited by A. Goldthau and J. M. Witte. Washington, DC: Brookings Institution Press.

Kohler-Koch, Beate and Berthold Rittberger. 2006. The 'governance turn' in EU studies. *Journal of Common Market Studies* 44:27–49.

Konoplyanik, Andrey A. 2010. Evolution of gas pricing in Continental Europe: a view from Russia – modernization of indexation formulas versus gas-to-gas competition. *Oil, Gas & Energy Law Intelligence*. June.

Kovacevic, Aleksandar. 2009. The Impact of the Russia–Ukraine Gas Crisis in South Eastern Europe. Oxford Institute for Energy Studies. NG 29.

KPMG. 2012. Central and Eastern European Shale Gas Outlook. KPMG Global Energy Institute.

KPMG. 2014. New Shale Gas Tax Law in Poland. Warsaw.

Krasner, Stephen D. 1983. Structural causes and regime consequences: regimes as intervening variables. In *International Regimes*, edited by S. D. Krasner. Ithaca, NY: Cornell University Press.

Krastanova, Radosveta. 2012. *The Green Movement and the Green Parties in Bulgaria: Between System Integration and System Change*. Sofia: Friedrich Ebert Foundation

Krastev, Ivan. 2014. Russian revisionism. Putin's plan for overturning the European order. *Foreign Affairs*. 3 March. Available from https://www.foreignaffairs.com/articles/russia-fsu/2014-03-03/russian-revisionism [accessed 12 June 2017].

Kriesky, Jill, Bernard D. Goldstein, Katrina Zell and Scott Beach. 2013. Differing opinions about natural gas drilling in two adjacent counties with different levels of drilling activity. *Energy Policy* 58:228–236.

Kuhn, Maximilian and Frank Umbach. 2011. Unconventional gas resources: a transatlantic shale alliance? In *Transatlantic Energy Futures. Strategic Perspectives on Energy Security, Climate Change and New Technologies in Europe and the United States*, edited by D. Koranyi. Washington, DC: Center for Transatlantic Relations, Paul H. Nitze School of Advanced International Studies, John Hopkins University Press, pp. 207–228.

Kuzemko, Caroline, Andrei Belyi, Andreas Goldthau and Michael Keating, eds. 2012. *Dynamics of Energy Governance in Europe and Russia*. Basingstoke: Palgrave Macmillan.

LaBelle, Michael. 2017. A state of fracking: building Poland's national innovation capacity for shale gas. *Energy Research & Social Science* 23:26–35.

LaBelle, Michael C. 2016. Failure to frack: pitfalls of governance and risk in Polish shale gas. In *Risks, Rewards and Regulation of Unconventional Gas. A Global Perspective*, edited by R. Q. Grafton, I. G. Cronshaw and M. C. Moore. Cambridge: Cambridge University Press, pp. 267–285.

LaBelle, Michael and Andreas Goldthau. 2014a. Escaping the valley of death? Comparing shale gas technology policy prospects to nuclear and solar in Europe. *Journal of World Energy, Law & Business* 7 (2):93–111.

LaBelle, Michael and Andreas Goldthau. 2014b. The governance of shale gas in Bulgaria. From exploration to bust. *Oil, Gas and Energy Law Journal* 12 (3): Special Edition on The Governance of Unconventional Gas Development Outside the United States.

Lachapelle, Erick, Éric Montpetit and Jean-Philippe Gauvin. 2014. Public perceptions of expert credibility on policy issues: the role of expert framing and political worldviews. *Policy Studies Journal* 42 (4):674–697.

Landesamt für Bergbau, Energie und Geologie. 2013. Erdöl und Erdgas in der Bundesrepublic Deutschland. Hannover.

Landry, David. 2013. Fracking for Hungary? Minister describes Hungary's 'neutral' position. *Budapest Business Journal*. 19 July.

Larsson, Robert L. 2006. *Russia's Energy Policy: Security Dimensions and Russia's Reliability as an Energy Supplier*. Stockholm: Swedish Defence Research Agency.

Lauria, Mickey. 1996. Reconstructing Urban Regime Theory. In *Reconstructing Urban Regime Theory: Regulating Urban Politics in a Global Economy*, edited by M. Lauria. Thousand Oaks, CA: Sage, 1–10.

Law, John. 1991. Theory and narrative in the history of technology: response. *Technology and Culture* 32 (2):379–380.

Levi-Faur, David. 1999. The Governance of Competition: the interplay of technology, economics, and politics in European Union electricity and telecom regimes. *Journal of Public Policy* 19 (2):175–207.

Lieberman, Robert C. 2002. Ideas, institutions, and political order: explaining political change. *American Political Science Review* 96 (4):697–712.

Lin, Albert C. 2014. Fracking and federalism: a comparative approach to reconciling national and subnational interests in the United States and Spain *Environmental Law* 44 (4):1039–1078

Lipsky, Michael. 1980. *Street-level Bureaucracy: Dilemmas of the Individual in Public Services*. New York, NY: Russell Sage Foundation.

Lis, Aleksandra and Piotr Stankiewicz. 2016. Framing shale gas for policy-making in Poland. *Journal of Environmental Policy & Planning* 19 (1):53–71.

Lis, Aleksandra, Claudia Braendle, Torsten Fleischer, et al. 2015. Existing European Data on Public Perceptions of Shale Gas. M4ShaleGas Consortium. Project Report.

Lodge, Martin. 2002. Varieties of Europeanisation and the national regulatory state. *Public Policy and Administration* 17 (2):43–67.

Lodge, Martin. 2008. Regulation, the regulatory state and European politics. *West European Politics* 31 (1/2):280–301.

Lowi, Theodore. 1964. American Business, Public Policy, Case-Studies, and Political Theory. *World Politics* 16:677–715.

Lund, Susan, James Manyika, Scott Nyquist, Lenny Mendonca and Sreenivas Ramaswamy. 2013. *Game Changers: Five Opportunities for US Growth and Renewal*, Seoul/San Francisco: McKinsey Global Institute.

Majone, Giandomenico. 1994. The rise of the regulatory state in Europe. *West European Politics* 17 (3):77–101.

Majone, Giandomenico. 1997. From the positive to the regulatory state: causes and consequences of changes in the mode of governance. *Journal of Public Policy* 17 (2):139–167.

Manley, David, James Cust and Giorgia Cecchinato. 2017. Stranded Nations? The Climate Policy Implications for Fossil Fuel-Rich Developing Countries. Oxford: Oxford Centre for the Analysis of Resource Rich Economies. OxCarre Policy Paper 34.

Mason, Charles F., Lucija A. Muehlenbachs and Sheila M. Olmstead. 2015. *The Economics of Shale Gas Development*. Washington, DC: Resources for the Future.

May, Peter J. and Ashley E. Jochim. 2013. Policy regime perspectives: policies, politics, and governing. *Policy Studies Journal* 41 (3):426–452.

May, Peter J., Ashley E. Jochim and Joshua Sapotichne. 2011. Constructing homeland security: an anemic policy regime. *Policy Studies Journal* 39 (2):285–307.

Mazmanian, Daniel A. and Paul A. Sabatier. 1989. *Implementation and Public Policy*. Lanham, MD: University Press of America.

McCool, Daniel. 1998. The subsystem family of concepts: a critique and a proposal. *Political Research Quarterly* 51 (2):551–570.

McCubbins, Mathew D., Roger G. Noll and Barry R. Weingast. 1987. Administrative procedures as instruments of political control. *Journal of Law, Economics, & Organization* 3 (2):243–277.

McCubbins, Mathew D., Roger G. Noll and Barry R. Weingast. 1989. Structure and process, politics and policy: administrative arrangements and the political control of agencies. *Virginia Law Review* 75 (2):431–482.

McGowan, Francis. 2014. Regulating innovation: European responses to shale gas development. *Environmental Politics* 23 (1):41–58.

McGowan, Francis and Helen Wallace. 1996. Towards a European regulatory state. *Journal of European Public Policy* 3 (4):560–576.

McGuinn, Patrick. 2006. Swing issues and policy regimes: federal education policy and the politics of policy change. *Journal of Policy History* 18 (2):205–240.

Medarov, Georgi. 2013. Land concentration, land grabbing and land conflicts in Europe: the case of Boynitsa in Bulgaria. In *Land Concentration, Land Grabbing and People's Struggles in Europe*, edited by J. Franco and S. M. Borras Jr. Amsterdam: Transnational Institute (TNI)/European Coordination Via Campesina (ECVC)/Hands off the Land Alliance, pp. 154–179.

Medlock, Kenneth B. 2009. *Shale Gas: A Game-Changer with Global Implications*. Houston, TX: James A. Baker III Institute for Public Policy of Rice University.

Medlock, Kenneth B., Amy Myers Jaffe and Peter R. Hartley. 2011. *Shale Gas and US National Security*. Houston, TX: James A. Baker III Institute for Public Policy.

Meier, Kenneth J. and Kevin B. Smith. 1994. Say it ain't so, Moe: institutional design, policy effectiveness, and drug policy. *Journal of Public Administration and Theory* 4 (5):429–442.

Metze, Tamara. 2017. Fracking the debate: frame shifts and boundary work in Dutch decision making on shale gas. *Journal of Environmental Policy & Planning* 19 (1):35–52.

Mihalache, Anca Elena. 2015. No shale gas, after all – implications of Chevron's exit from Romania. *Natural Gas Europe*. 30 March.

Milward, H. Brinton and Keith G. Provan. 2000. Governing the hollow state. *Journal of Public Administration Research and Theory* 10 (2):359–380.

Ministerstwo Gospodarki/Ministerstwo Srodowiska. 2014. Strategia Bezpieczeństwo Energetyczne i Środowisko perspektywa do 2020 r. Warszawa.

Ministerul Energiei. 2016. Strategia Energetică a României 2016–2030, cu perspectiva anului 2050. Bukarest

Ministry of Economics, Energy and Tourism. 2011. Energy Strategy of Republic of Bulgaria till 2020. For Reliable, Effective and Cleaner Energy. Sofia

Ministry of Economy and Energy. 2014. Energy Security and the Environment Strategy – 2020 Perspective (Bezpieczeństwo energetyczne i środowisko – perspektywa do roku 2020). Warsaw.

Mossberger, Karen and Gerry Stoker. 2001. The evolution of urban regime theory: the challenge of conceptualization. *Urban Affairs Review* 36 (6):810–835.

Murphy, Joseph, ed. 2007. *Governing Technology for Sustainability*. London: Routledge.

Natural Gas Europe. 2012. Park Place Energy Not Seeking to Explore for Shale in Bulgaria. 5 April.

Natural Gas Europe. 2013a. Bureaucratic Delays Impact Poland's Shale Promise. 1 February.

Natural Gas Europe. 2013b. Poland Proposes Restrictions to Shale Gas Opposition. 1 April.

Naturalgasworld. 2012. President: Romania Needs Decision on Shale Gas Development. 13 August.

Naturalgasworld. 2013. Romania Needs Shale Gas to Achieve Energy Independence. 6 October.

Nelsen, Jacqueline L. 2006. Social license to operate. *International Journal of Mining, Reclamation and Environment* 20 (3):161–162.

Neslen, Arthur. 2014. EC serves notice to Poland over shale gas defiance. *The Guardian*. 30 July.

Neville, Kate J. and Erika Weinthal. 2016. Scaling up site disputes: strategies to redefine 'local' in the fight against fracking. *Environmental Politics* 25 (4):569–592.

Neville, Kate J., Jennifer Baka, Shanti Gamper-Rabindran, et al. 2017. Debating unconventional energy: social, political, and economic implications. *Annual Review of Environment & Resources* 42. Available from: https://doi.org/10.1146/annurev-environ-102016–061102 [accessed 22 November 2017].

Newman, Joshua and Michael Howlett. 2014. Regulation and time: temporal patterns in regulatory development. *International Review of Administrative Sciences* 80 (3):493–511.

Newsweek. 2015. Romania's Prime Minister appears in court days after shock resignation. 11 June.

New York Times. 1981. Soviet natural gas for West curtailed. 10 January.

New York Times. 2010. Eastern Europe, seeking energy security, turns to shale gas. 11 May.

New York Times. 2012a. Romanian Prime Minister resigns amid protests. 6 February.

New York Times. 2012b. Seeking disclosure on fracking. 30 May.

New York Times. 2013. The lives they lived: George Mitchell. 21 December.

New York Times. 2014a. Boom in energy spurs industry in the Rust Belt. 8 September.

New York Times. 2014b. Lithuania offers example of how to break Russia's grip on energy. 27 October.

New York Times. 2017. Anger and mistrust fuel unabated protests in Romania. 12 February.

Nikolenyi, Csaba. 2014. *Institutional Design and Party Government in Post-Communist Europe*. Oxford: Oxford University Press.

North, Douglass. 1990. *Institutions, Institutional Change and Economic Performance*. New York, NY: Cambridge University Press.

North, D. Warner, Paul C. Stern, Thomas Webler and Patrick Field. 2014. Public and stakeholder participation for managing and reducing the risks of shale gas development. *Environmental Science & Technology* 48 (15):8388–8396.

Nosko, Andrej and Matúš Mišík. 2017. No united front: the political economy of energy in Central and Eastern Europe. In *Energy Union. Europe's New Liberal Mercantilism?*, edited by S. Andersen, A. Goldthau and N. Sitter. Basingstoke: Palgrave Macmillan, pp. 201–222.

Novinite. 2011a. Bulgarian minister: energy is key to national security. 29 May.

Novinite. 2011b. Dundee: Bulgaria's gold mine project has 3.3 year payback. 6 December.

Novinite. 2012. Bulgaria quits belene nuclear power plant project. 28 March.

Novinite. 2015. Bulgaria's energy sector is in state of financial collapse – minister. 22 February.

Nülle, Grant Mark. 2015. Prospects for shale development outside the USA: evaluating nations' regulatory and fiscal regimes for unconventional hydrocarbons. *Journal of World Energy Law & Business* 8 (3):232–268.

Nygren, Bertil. 2008. Putin's use of natural gas to reintegrate the CIS region. *Problems of Post-Communism* 55 (4):3–15.

Oberthür, Sebastian and Olav Schram Stokke, eds. 2011. *Managing Institutional Complexity. Regime Interplay and Global Environmental Change*. Cambridge, MA: MIT Press.

Orban, Anita. 2008. *Power, Energy, and the New Russian Imperialism*. Santa Barbara, CA: Praeger.

Orren, Karen and Stephen Skowronek. 1998. Regimes and regime building in American government: a review of literature on the 1940s. *Political Science Quarterly* 113 (4):689–702.

O'Sullivan, Meghan. 2013. The entanglement of energy, grand strategy, and international security. In *Wiley Handbook of Global Energy Policy*, edited by A. Goldthau. London: Wiley-Blackwell.

Owen, John R. and Deanna Kemp. 2013. Social licence and mining: a critical perspective. *Resources Policy* 38 (1):29–35.

Pachiu, Laurentiu, Raluca Mustaciosu and Radu Dudau. 2014. Oil and Gas Regulation in Romania: An Overview. Energy and Natural Resources Multi-Jurisdictional Guide.

Pachiu, Laurentiu, Delia Vasiliu and Radu Dudău. 2016. Romania. In *Energy 2017*, 5th edn, edited by G. Picton-Turbervill and J. Derrick. London: Global Legal Group Ltd, pp. 180–186.

Parsons, Richard and Kieren Moffat. 2014. Constructing the meaning of social licence. *Social Epistemology* 28 (3–4):340–363.

Pearson, Ivan, Peter Zeniewski, Francesco Gracceva, et al. 2012. *Unconventional Gas: Potential Energy Market Impacts in the European Union.* Luxembourg: European Commission Joint Research Centre.

Pencea, Roxana, Tudor Brădăţan and Ştefania Simion. 2013. *Transylvania. Undermined-Territory*. Bukarest: Mining Watch România.

Peters, B. Guy. 1998. Managing horizontal government: the politics of co-ordination. *Public Administration* 76 (2):295–311.

Peters, Guy. 2010. Bureaucracy and democracy. *Public Organization Review* 10:209–222.

PGI-NRI. 2012. Assessment of Shale Gas and Shale Oil Resources of the Lower Paleozoic Baltic-Podlasie-Lublin Basin in Poland, Warsaw. Polish Geological Institute – National Research Institute. Available from https://www.pgi.gov.pl/docman-tree/aktualnosci-2012/zasoby-gazu/769-raport-en/file.html [accessed 22 November 2017].

Pickles, John and Adrian Smith, eds. 1998. *Theorizing Transition: The Political Economy of Post-Communist Transformations*. London: Routledge.

Pinch, Trevor J. and Wiebe Bijker. 1984. The social construction of facts and artifacts: or how the sociology of science and the sociology of technology might benefit each other. *Social Studies of Science* 14 (3):399–418.

PKEE. 2016. Polish Power Sector. Getting the Facts Straight. Warsaw: Polish Electricity Association.

Platts. 2017. Analysis: Global Gas Suppliers Up The Ante in Market Share Battle. 24 February.

Polish Geological Survey. 2014. Public Opinion Poll on Shale Gas. Available from http://infolupki.pgi.gov.pl/en/society/public-opinion-poll-shale-gas [accessed 14 July 2016].

Polish Geological Survey. 2015a. Shale Gas Exploration Status in Poland as of April 2015. Available from http://infolupki.pgi.gov.pl/en/exploration-status/news/shale-gas-exploration-status-poland-april-2015 [accessed 1 July 2015].

Polish Geological Survey. 2015b. Taxation of Hydrocarbons – Not Really High – Will Have No Adverse Effects on the Level of Investment.

Available from http://infolupki.pgi.gov.pl/en/economy/law-concessions/taxation-hydrocarbons-not-really-high-will-have-no-adverse-effects-level [accessed 22 November 2017].

Polish Ministry of the Environment. 2015. Shale Gas Development in Poland – Progress Report.

Politico. 2013. Shale gas could be 'game changer'. 24 April.

Politico. 2016. Shrugging off concerns, Europe waves through Hungary's controversial nuke deal. 27 December.

Pollitt, Christopher. 2003. Joined-up government: a survey. *Political Studies Review* 1 (1):34–49.

Pöyry. 2013. Macroeconomic Effects of European Shale Gas Production. A Report to the International Association of Oil and Gas Producers (OGP). Oxford.

PricewaterhouseCoopers. 2011. Shale Gas: A Renaissance in US Manufacturing. New York.

The PRS Group. 2015. Regional Political Risk Index (as of April 2015). Available from https://www.prsgroup.com/category/risk-index [accessed 7 October 2016].

Prno, Jason and D. Scott Slocombe. 2012. Exploring the origins of 'social license to operate' in the mining sector: Perspectives from governance and sustainability theories. *Resources Policy* 37:346–357.

Pye, Steve, Audrey Dobbins, Claire Baffert, et al. 2015. Energy Poverty and Vulnerable Consumers in the Energy Sector across the EU: Analysis of Policies and Measures. Brussels: INSIGHT_E.Policy Report 2.

Rabe, Barry G. and Christopher Borick. 2013. Conventional politics for unconventional drilling? Lessons from Pennsylvania's early move into fracking policy development. *Review of Policy Research* 30 (3):321–340.

Rahm, Dianne. 2011. Regulating hydraulic fracturing in shale gas plays: the case of Texas. *Energy Policy* 39 (5):2974–2981.

Rayner, Jeremy and Michael Howlett. 2009. Introduction: understanding integrated policy strategies and their evolution. *Policy and Society* 28 (2):99–109.

Rein, Martin. 1983. *From Policy to Practice*. Harmondsworth: Penguin.

Reins, Leonie. 2017. *Regulating Shale Gas. The Challenge of Coherent Environmental and Energy Regulation*. Cheltenham: Edward Elgar.

Reuters. 2010. Chevron seeks to explore for shale gas in Bulgaria. 15 July.

Reuters. 2011. Bulgaria seeks to ease fears on shale gas drilling. 19 July.

Reuters. 2012a. Bulgaria eases oil and gas fracking ban. 14 June.

Reuters. 2012b. Czechs eye moratorium on shale gas exploration. 7 May.

Reuters. 2013a. Exit by two foreign firms leaves Polish shale gas under cloud. 8 May.

Reuters. 2013b. Romanian farmers choose subsistence over shale gas. 19 October.

Reuters. 2013c. Work together better, shale gas investors tell Polish companies. 27 March.

Reuters. 2014a. Chevron leaves Lithuania as shale gas prospects remain uncertain. 8 July.

Reuters. 2014b. Romania does not have shale gas, PM Ponta says. 10 November.

Reuters. 2014c. US considers fracking fluid disclosure rules. 9 May.

Reuters. 2015a. Bulgaria scraps incentives for new renewable energy installations. 26 February.

Reuters. 2015b. Conoco the last global oil firm to quit Polish shale gas. 5 June.

Reuters. 2015c. Insight – Creeping 'Orbanisation' in Poland sparking unease in EU. 17 December.

Reuters. 2015d. Update 1 – Romania to send oil and gas royalty tax law to parliament in September. 25 May.

Reuters. 2016a. Dutch lower Groningen gas production cap to 24 bcm. 24 June.

Reuters. 2016b. Poland aims to end long-term gas supplies from Russia after 2022. 31 May.

Rhodes, R.A.W. 1997. *Understanding Governance. Policy Networks, Governance, Reflexivity and Accountability*. Buckingham: Open University Press.

Rittel, Horst W. J. and Melvin M. Webber. 1973. Dilemmas in a general theory of planning. *Policy Sciences* 4:155–169.

Rodrik, Dani, Arvind Subramanian and Francesco Trebbi. 2004. Institutions rule: the primacy of institutions over geography and integration in economic development. *Journal of Economic Growth* 9 (2):131–165.

Romanian Energy Regulatory Authority. 2015a. First Quarter 2015 Newsletter. Bucharest: ANRE.

Romanian Energy Regulatory Authority. 2015b. National Report 2014. Bucharest: ANRE.

Romanian National Committee of World Energy Council. 2013. Natural Gas Resource from Unconventional Fields – Potential and Recovery. Bucharest.

Rothstein, Bo and Jan Teorell. 2008. What is quality of Government? A theory of impartial government institutions. *Governance* 21 (2):165–190.

Rutland, Peter. 2008. Russia as an energy superpower. *New Political Economy* 13 (2): 203–210.

Sabatier, P. A. 1988. An advocacy coalition framework of policy change and the role of policy-oriented learning therein. *Policy Sciences* 21:129.

Sabatier, Paul A. 1986. Top-down and bottom-up approaches to implementation research: a critical analysis and suggested synthesis. *Journal of Public Policy* 6 (1):21–48.

Sabatier, Paul A. and Hank C. Jenkins-Smith. 1999. The advocacy coalition framework: an assessment. In *Theories of the Policy Process*, edited by P. A. Sabatier. Boulder, CO: Westview Press.

Sabel, Charles F. and Jonathan Zeitlin, eds. 2010. *Experimentalist Governance in the European Union. Towards a New Architecture.* Oxford: Oxford University Press.

Sabel, Charles F. and Jonathan Zeitlin. 2012. Experimentalist governance. In *The Oxford Handbook of Governance*, edited by D. Levi-Faur. Oxford: Oxford University Press, pp. 169–183.

Sakmar, Susan L. 2011. The global shale gas initiative: will the United States be the role model for the development of shale gas around the world? *Houston Journal of International Law* 33 (2):369–416.

Schafft, Kai A., Yetkin Borlu and Leland Glenna. 2013. The relationship between marcellus shale gas development in Pennsylvania and local perceptions of risk and opportunity the relationship between Marcellus Shale gas development in Pennsylvania and local perceptions of risk and opportunity. *Rural Sociology* 78 (2):143–166.

Scharpf, Fritz W. 1999. *Governing in Europe: Effective and Democratic.* Oxford: Oxford University Press.

Schattschneider, Elmer Eric. 1935. *Politics, Pressure, and the Tariff.* New York, NY: Prentice-Hall.

Schmidt, Vivien A. 2013. Democracy and legitimacy in the European Union revisited: input, output and 'throughput'. *Political Studies* 61 (1):2–22.

Schmitt, Sophie. 2013. Comparative approaches to the study of public policy-making. In *Routledge Handbook of Public Policy*, edited by E. J. Araral, S. Fritzen, M. Howlett, M. Ramesh and X. Wu. New York, NY: Routledge, pp. 29–43.

Schneider, Anne L. and Helen M. Ingram. 2009. What is next for policy design and social construction theory? *Policy Studies Journal* 37 (1):103–119.

Shaffer, Brenda. 2009. *Energy Politics.* Philadelphia, PA: University of Pennsylvania Press.

Silve, Florent and Pierre Noël. 2010. Cost curves for Gas Supply Security: the Case of Bulgaria. Cambridge Electricity Research Group Working Paper 1031.

Simina, Anca. 2013. Ponta susţine nu doar explorarea, ci şi exploatarea gazelor de şist: 'Acum nu mai am dubii'. *Gandul.* 13 June.

Skalamera, Morena and Andreas Goldthau. 2016. Russia Playing Hardball or Bidding Farewell to Europe? Debunking the Myths of Eurasia's New Geopolitics of Gas. Harvard White Paper.

Smeenk, Tom. 2010. *Russian Gas for Europe: Creating Access and Choice. Underpinning Russia's Gas Export Strategy with Gazprom's Infrastructure Investments*. The Hague: Clingendael International Energy Programme.

Smil, Vaclav. 2010. *Energy Transitions: History, Requirements, Prospects*. Santa Barbara, CA: Praeger.

Smith, Don C. and Jessica M. Richards. 2015. Social license to operate: hydraulic fracturing-related challenges facing the oil & gas industry. *Oil and Gas, Natural Resources & Energy Journal* 1 (2):81–163.

Smith, Michael F. and Denise P. Ferguson. 2013. 'Fracking democracy': issue management and locus of policy decision-making in the Marcellus Shale gas drilling debate. *Public Relations Review* 39 (4):377–386.

The Sofia Echo. 2012. '75% of Bulgarians favour shale gas exploration if environmental safety certain' – poll. 17 January.

Sovacool, Benjamin K. 2014a. Cornucopia or curse? Reviewing the costs and benefits of shale gas hydraulic fracturing (fracking). *Renewable & Sustainable Energy Reviews* 37 (September):249–264.

Sovacool, Benjamin K. 2014b. What are we doing here? Analyzing fifteen years of energy scholarship and proposing a social science research agenda. *Energy Research & Social Science* 1:1–29.

Spence, David B. 2013. Federalism, regulatory lags, and the political economy of energy production. *University of Pennsylvania Law Review* 161 (431):431–508.

Standard News. 2014. US shale gas giant Chevron is leaving Bulgaria. 28 May.

Stark, David and Laszlo Bruszt. 1998. *Postsocialist Pathways: Transforming Politics and Property in East Central Europe*. Cambridge: Cambridge University Press.

State Agency for National Security. 2011. National Security Strategy of the Republic of Bulgaria. Sofia.

Stefanov, Ruslan, Valentina Nikolova, Dobromir Hristov, et al. 2011. *Energy and Good Governance in Bulgaria. Trends and Policy Options*. Sofia: Center for the Study of Democracy

Stefanov, Ruslan, Todor Galev, Martin Tsanov, Martin Vladimirov and Nadejda Gantcheva. 2014. *Energy Sector Governance and Energy (In)security in Bulgaria*. Sofia: Center for the Study of Democracy.

Stein, Robert M. and Kenneth N. Bickers. 1995. *Perpetuating The Pork Barrel: Policy Subsystems and American Democracy*. Cambridge: Cambridge University Press.

Stent, Angela. 1981. *From Embargo to Ostpolitik: The Political Economy of West German-Soviet Relations, 1955–1980*. Cambridge: Cambridge University Press.

Stephenson, Michael H. 2016. Shale gas in North America and Europe. *Energy Science and Engineering* 4 (1):4–13.

Stern, Jonathan. 2006. *The Russian–Ukrainian Gas Crisis of 2006*. Oxford: Oxford Institute for Energy Studies.

Stern, Jonathan, Simon Pirani and Katja Yafimava. 2009. *The Russo-Ukrainian Gas Dispute of January 2009: A Comprehensive Assessment*. Oxford: Oxford Institute for Energy Studies.

Stevens, Paul. 2009. *Transit Troubles. Pipelines as a Source of Conflict*. London: Chatham House.

Stone, Clarence N. 1989a. *Regime Politics: Governing Atlanta 1946–1988*. Lawrence, KS: University Press of Kansas.

Stone, Deborah A. 1989b. Causal stories and the formation of policy agendas. *Political Science Quarterly* 104 (2):281–300.

Stone, Deborah. 2002. *Policy Paradox: The Art of Political Decision Making*. New York, NY: W. W. Norton.

Stulberg, Adam N. 2005. Moving beyond the great game: the geoeconomics of Russia's influence in the Caspian energy bonanza. *Geopolitics* 10 (1):1–25.

Stulberg, Adam N. 2008. *Well-Oiled Diplomacy: Strategic Manipulation and Russia's Energy Statecraft in Eurasia*. New York, NY: SUNY Press.

Stulberg, Adam N. 2012. Strategic bargaining and pipeline politics: confronting the credible commitment problem in Eurasian energy transit. *Review of International Political Economy* 19 (5):808–836.

Supreme Audit Office of Poland. 2014. NIK on Shale Gas Search – Supreme Audit Office. Warsaw.

Szulecki, Kacper, Severin Fischer, Anne Therese Gullberg and Oliver Sartor. 2016. Shaping the 'Energy Union': between national positions and governance innovation in EU energy and climate policy. *Climate Policy* 16 (5):548–567.

Thomson, Ian and Robert Boutilier. 2011. The social licence to operate. In *SME Mining Engineering Handbook*, edited by P. Darling. Englewood, CO: Society for Mining, Metallurgy, and Exploration, pp. 1779–1796.

Trammel, Setephan. 2015. The Unconventional is Becoming the New Conventional. Available from http://blog.ihs.com/the-unconventional-is-becoming-the-new-conventional [accessed 22 May 2017].

Transgaz. 2016. Structure of Gas Mixture. Available from www.transgaz.ro/en/informatii-clienti-0 [accessed 22 July 2016].

Transparency International. 2014. Corruption Perception Index 2014: Results. Berlin.

Tusk, Donald. 2014. A united Europe can end Russia's energy stranglehold. *Financial Times*. 21 April.

Ulli-Beer, Silvia. 2013. *Dynamic Governance of Energy Technology Change: Socio-technical Transitions towards Sustainability*. Heidelberg: Springer.

Umbach, Frank. 2013. The unconventional gas revolution and the prospects for Europe and Asia. *Asia Europe Journal* 11 (3):305.

Upham, Paul, Aleksandra Lis, Hauke Riesch and Piotr Stankiewicz. 2015. Addressing social representations in socio-technical transitions with the case of shale gas. *Environmental Innovation and Societal Transitions* 16:120–141.

UPI. 2013. Romgaz lands option for Black Sea field. 18 February.

Urban, Julie A. 2008. US access to the global LNG market. *OPEC Energy Review* 32 (3):215–231.

US Bureau of Labor Statistics. 2014. The Marcellus Shale Gas Boom in Pennsylvania: Employment and Wage Trends. Monthly Labor Review. Available from https://doi.org/10.21916/mlr.2014.7 [accessed 6 June 2017].

Van de Graaf, Thijs, Tim Haesebrouck and Peter Debaere. 2017. Fractured politics? The comparative regulation of shale gas in Europe. *Journal of European Public Policy*:1–18.

Verweij, Marco, Mary Douglas, Richard Ellis, et al. 2006. Clumsy solutions for a complex world: the case of climate change. *Public Administration* 84 (4):817–843.

Vesalon, Lucian and Remus Cretan. 2015. 'We are not the Wild West': anti-fracking protests in Romania. *Environmental Politics* 24 (2):288–307.

Visan, George. 2013. Public Perception of shale gas in Romania. Bucarest: Romania Energy Center. Policy Brief 3.

VOA. 2015. Bulgaria key battleground in US–Russia energy war. 23 February.

Wallace, Helen, Mark A. Pollack and Alasdair R. Young, eds. 2014. *Policy-Making in the European Union*, 7th edn. Oxford: Oxford University Press.

Wall Street Journal. 2012. Polish state-controlled firms team up to explore for shale gas. 4 July.

Wall Street Journal. 2014. Poland proposes tax breaks for shale gas industry. 11 March.

Wampler, Brian. 2007. *Participatory Budgeting in Brazil: Contestation, Cooperation, and Accountability*. Philadelphia, PA: Penn State University Press.

Wang, Zhongmin and Alan Krupnick. 2013. *A Retrospective Review of Shale Gas Development in the United States*. Washington, DC: Resources for the Future.

Warner, Barbara and Jennifer Shapiro. 2013. Fractured, fragmented federalism: a study in fracking regulatory policy. *Publius: The Journal of Federalism* 43 (3):474–496.

Watson, Nigel, Hugh Deeming and Raphael Treffny. 2009. Beyond bureaucracy? Assessing institutional change in the governance of water in England. *Water Alternatives* 2 (3):448–460.

Weaver, Kent. 2010. Paths and forks or chutes and ladders? Negative feedbacks and policy regime change. *Journal of Public Policy* 30 (2):137–162.

Weible, Christopher M., Tanya Heikkila, Karin Ingold and Manuel Fischer, eds. 2016. *Policy Debates on Hydraulic Fracturing. Comparing Coalition Politics in North America and Europe*. Basingstoke: Palgrave Macmillan.

Wendt, Alexander. 1994. Collective identity formation and the international state. *American Political Science Review* 88:384–96.

Wilber, Tom. 2012. *Under the Surface: Fracking, Fortunes, and the Fate of the Marcellus Shale*. Ithaca, NY: Cornell University Press.

Williams, Laurence, Phil Macnaghten, Richard Davies and Sarah Curtis. 2015. Framing 'fracking': Exploring public perceptions of hydraulic fracturing in the United Kingdom. *Public Understanding of Science* 26(1):89–104.

Williams, Russell A. 2009. Exogenous shocks in subsystem adjustment and policy change: the credit crunch and Canadian banking regulation. *Journal of Public Policy* 29 (1):29–53.

Wilson, Carter. 2000. Policy regimes and policy change. *Journal of Public Policy* 20 (3):247–274.

Wolsink, Maarten. 2000. Wind power and the NIMBY-myth: institutional capacity and the limited significance of public support. *Renewable Energy* 21 (1):49–64.

Wolske, Kim, Andrew Hoffman and Lukas Strickland. 2013. *Public Perceptions of High-Volume Hydraulic Fracturing & Deep Shale Gas Development*. Graham Sustainability Institute Integrated Assessment Report Series. Ann Arbor, MI: University of Michigan, pp. 8.

Worsham, Jeff and Chaun Stores. 2012. Pet sounds: subsystems, regimes, policy punctuations, and the neglect of African American farmers, 1935–2006. *Policy Studies Journal* 40 (1):169–189.

Youngs, Richard. 2009. *Energy security: Europe's New Foreign Policy Challenge*. London: Routledge.

Zafonte, Matthew and Paul A. Sabatier. 1998. Share beliefs and imposed interdependencies as determinants of ally networks in overlapping subsystems. *Journal of Theoretical Politics* 10 (4):473–505.

Zborowski, Matt. 2015. Poland's energy challenge. *Oil & Gas Journal*. 28 September.

Zielinski, Michal. 2012. Pomorze: seeds of polish shale gas counter revolution. *Natural Gas Europe*. 26 November.

Zuckerman, Gregory. 2013. *The Frackers: The Outrageous Inside Story of the New Billionaire Wildcatters*. New York, NY: Portfolio/Penguin.

Index

Act on a Special Hydrocarbon Tax, 6, 78, 81
Argentina, 2
Austria, 13, 23, 96, 134

Baia Mare, 134
ban. *See* moratorium
Bankwatch Network, 101, 102
Băsescu, Traian, 119, 130, 132
Basin Directorates (Bulgaria), 95
Birds Directive, 42
Black Sea, 17, 26, 94, 96, 116, 117, 120, 121, 124, 126, 127, 131
Boc, Emil, 119
Borisov, Boyko, 4, 99, 105, 112
Bulgargas, 24, 96, 97, 98, 100, 101, 103, 140
Bulgaria, 2, 4, 6, 10, 12, 13, 14, 16, 17, 24, 28, 29, 30, 36, 45, 66, 87, 91, 92, 93, 94, 95, 96, 97, 98, 99, 100, 102, 104, 106, 107, 108, 109, 110, 111, 112, 113, 114, 115, 116, 118, 120, 129, 132, 133, 136, 138, 139, 140, 141, 142, 143, 144, 145, 146, 147, 148, 150
Bulgaria Energy Holding, 96, 100, 101
Bulgaria–Romania–Hungary–Austria Interconnector, 121
Bulgartransgaz, 96, 100

Canada, 1, 2, 22, 31, 115, 124, 135
carbon footprint, 15, 21, 22, 86
CEE. *See* Central Eastern Europe
Central Eastern Europe, 2, 3, 4, 8, 25, 31, 33, 47, 71, 101, 102, 115, 116, 132
Chevron, 6, 13, 67, 87, 94, 96, 99, 101, 102, 107, 113, 115, 116, 124, 126, 127, 129, 131, 133, 139
China, 2, 11, 153
Citizens for European Development of Bulgaria, 105, 106, 113
Civic Platform, 70, 80
civil society organizations, 16, 74, 77, 83, 84, 87, 88, 90, 101, 103, 111, 124, 125, 128, 129, 135, 140, 143
Clean Drinking Water Act, 46
climate change, 39, 74
Climate Coalition, 74, 77, 81, 84, 88
Clinton, Hillary, 3
coal, 1, 3, 20, 21, 33, 40, 45, 69, 74, 83, 85, 86, 92, 99, 109, 121, 135
Cold War, 2
COMAH Directive, 41
COMECON, 23, 24
communism, 2, 15, 23, 47, 79, 122, 129
comparative public policy, 4, 6, 7, 10, 17, 48, 57, 153
concession, 68, 72, 73, 75, 76, 81, 82, 87, 96, 102, 106, 122, 123, 126, 127, 130, 131
Concessions Law (Bulgaria), 95
consumer surplus, 19
copper, 3, 98
Czech Republic, 12, 13, 29, 40

decarbonization, 70, 86, 109, 120, 143, 152
Directive Ensuring Equal Access to Exploration Licences and Authorizations, 42
Directive on Environmental Impact Assessments, 39, 41, 79
Directive on Registration, Evaluation, Authorization and Restriction of Chemical Substances, 41
Directorate for Environment and Waters (Bulgaria), 95, 96, 99

diversification, 8, 69, 83, 94, 98, 138, 151
Dobrich, 100, 111
downstream, 25, 122

E&P. *See* exploration and production
EIA. *See* Energy Information Agency
Electricity and Natural Gas Law (Romania), 122
Energy and Water Regulatory Commission (Bulgaria), 91, 95, 97, 101
energy governance, 2, 25, 47, 71, 75, 79, 96, 98, 123
energy independence, 1, 31, 105, 119, 132, 142
Energy Information Agency, 1, 2, 13, 18, 19, 21, 30, 31, 67, 93, 94, 115
Energy Policy Act, 54
energy poverty, 92
Energy Regulatory Authority (Romania), 117, 121, 123
Energy Roadmap 2050, 33
energy security, 3, 8, 10, 11, 22, 28, 29, 31, 32, 35, 42, 43, 57, 69, 83, 105, 116, 118, 131, 147, 150, 153
Energy Security and Environmental Strategy (Poland), 70, 73
Energy Strategy 2016–2030 (Romania), 121
Energy Union, 40, 42
Eni, 37
ENI, 34, 67
Environment Protection Act (Bulgaria), 95
Environmental Impact Assessment, 41, 73, 77, 78, 81, 95, 99, 123
environmental movements, 45, 74, 87, 88, 97, 98, 101, 110, 114, 140
environmental NGOs. *See* environmental movements
Environmental Protection Agency, 54
EPA. *See* Environmental Protection Agency
EU. *See* European Union
European Commission, 28, 35, 38, 40, 41, 43, 46, 79
European Economic Area, 33
European Parliament, 40
European Union, 3, 33, 38, 39, 97, 125
Executive Agency on Environmental Protection (Bulgaria), 95
experimentalist governance, 82, 142
exploration and production, 2, 6, 44, 47, 72, 122, 130
exploration licence, 68, 99
extractive industries, 3, 7, 15, 45, 65, 74, 135, 142, 143, 149
ExxonMobil, 67, 120, 124, 127

Federation of the Industrial Energy Consumers (Bulgaria), 97, 100, 106, 108
financial crisis, 2, 36, 119, 120, 131, 133
fracking, 1, 2, 3, 7, 9, 10, 12, 13, 15, 16, 17, 18, 42, 43, 45, 54, 56, 63, 65, 66, 75, 87, 95, 96, 99, 102, 103, 105, 108, 109, 110, 112, 113, 116, 126, 127, 128, 132, 139, 140, 143, 146, 147, 148, 150, 151, 152, 153
fracking fluids, 3, 45, 46, 110, 128
Friends of the Earth Bulgaria, 101

Galata gas field, 96
gas disputes. *See* Ukrainian–Russian gas crisis
Gas Exporting Countries Forum, 18
gas markets, 1, 6, 8, 29, 33, 35, 36, 37, 44, 67, 151
gas transit, 11, 25, 26, 28, 142
Gasland, 102, 128
gas-on-gas competition, 36, 37, 38, 94, 113
Gazprom, 13, 25, 26, 28, 29, 34, 35, 37, 72, 91, 94, 96, 101, 113, 114, 119, 124, 132, 151
Gaz-System (Poland), 72
GDF Suez, 37
GDOS. *See* General Directorate for Environmental Protection (Poland)
General Directorate for Environmental Protection (Poland), 73, 81
Genting Oil & Gas, 104
Geological and Mining Act (Poland), 72, 81
geology, 6, 18, 33, 43, 68, 110, 116, 127

geopolitics, 1, 8, 10, 11, 26, 27, 28, 29, 30, 31, 35, 83, 106, 113, 116, 119, 131, 138, 142, 151
gold, 3, 98, 108, 134, 143, 144
Greenpeace, 124
Groundwater Directive, 41
groundwater safety, 3, 45, 87, 110, 143

Habitat Directive, 41
Halliburton Loophole, 46
Henry Hub, 19, 20
Hungary, 12, 13, 24, 25, 28, 29, 115, 124
hydraulic fracturing. *See* fracking
Hydrocarbon Investment Act (Poland), 78
Hydrocarbons Licensing Directive, 41

Imex Oil Ltd, 124
import dependence, 2, 8, 15, 22, 24, 29, 69, 106, 117, 139
Industrial Emissions Directive, 41
information sharing, 81, 141, 145, 146, 149
infringement procedure, 36, 97
institutional quality, 55, 129, 145
Interconnector Greece–Bulgaria, 36, 121
international security, 8, 9, 11, 15, 22, 29, 93, 151

Khan Asparuh, 96
Krumovgrad, 98, 108

Law and Justice Party, 70, 80
Law on Energy (Bulgaria), 95
liberalization, 25, 35, 55, 125
licensing, 6, 41, 42, 59, 67, 73, 76, 77, 105, 107, 115, 116, 123, 124, 126, 127, 130, 139, 142, 144
liquefied natural gas, 13, 18, 20, 22, 28, 32, 36, 37, 67, 69, 70, 72, 121, 153
Lisbon Treaty, 39
Lithuania, 12, 13, 28, 29, 115
LNG. *See* liquefied natural gas
long-term contracts, 3, 25, 34, 35, 37, 47
LOTOS Group, 72
LTC. *See* long-term contracts

Melrose Resources plc, 96
mineral rights, 44
mining, 3, 7, 45, 65, 73, 79, 81, 99, 108, 128, 134, 142, 143, 144, 149, 150
Mining Waste Directive, 41
Ministry of Economy and Energy (Bulgaria), 95, 96, 98, 102, 104, 113
Ministry of Energy (Romania), 123
Ministry of Environment (Poland), 68, 73
Ministry of Environment (Romania), 131
Ministry of Environment and Water (Bulgaria), 95, 96, 98, 99, 104, 110
Ministry of Foreign Affairs (Poland), 73, 83, 85
Ministry of the Economy (Poland), 73
Mitchell, George, 18, 43
MOL, 25, 115, 124
moratorium, 2, 7, 17, 45, 94, 113, 116, 126, 139, 150

nascent industry, 3, 12, 71, 78, 99
National Agency for Environmental Protection (Romania), 123, 128, 131
National Agency for Mineral Resources (Romania), 123, 128, 129, 131, 134
National Association of Grain Producers (Bulgaria), 111
National Operator of Energy Minerals (Poland), 80
Natura 2000, 78
natural gas, 1, 3, 5, 8, 11, 13, 18, 19, 20, 21, 25, 33, 40, 44, 46, 59, 69, 73, 79, 85, 86, 91, 93, 95, 96, 113, 120, 124, 125
NGOs. *See* civil society organizations
NIMBY, 54, 85
Nord Stream, 28
Norway, 33, 34, 37, 69, 84

offshore exploration, 2
Offshore Oil and Gas Directive, 42

oil, 3, 10, 19, 20, 22, 23, 25, 29, 31, 32, 35, 37, 38, 44, 45, 54, 55, 67, 72, 74, 76, 78, 80, 85, 86, 96, 102, 104, 106, 115, 122, 123, 128, 130, 136, 150
Oil and Gas Exploration and Production plc, 103, 104
Oil and Gas Safety Directive, 41
oil-price peg, 3
OMV, 96, 134
organizational arrangement, 4, 5, 11, 51, 53, 74, 89, 112, 137, 140, 141, 142, 145. *See also* policy regime
Orthodox Church (Romania), 125, 134, 135, 137, 144
Ostpolitik, 23

participatory governance, 5, 17, 53, 62, 63, 105, 127, 136, 138, 140, 142, 145, 146, 149
PERN, 72
Petroleum Law (Romania), 12, 122
Petrom, 120, 124, 125, 127, 134, 140
PGNiG, 24, 44, 72, 79, 80
PiS. *See* Law and Justice Party
PKN Orlen, 72, 76, 77, 80, 84
PKPP Lewiatan, 74, 84, 85, 87
PO. *See* Civic Platform
Poland, 2, 4, 6, 10, 12, 13, 14, 16, 17, 24, 28, 29, 30, 40, 41, 45, 66, 67, 68, 69, 70, 71, 72, 74, 75, 76, 77, 78, 80, 81, 82, 83, 84, 85, 86, 87, 88, 90, 94, 115, 116, 118, 129, 138, 139, 140, 141, 142, 143, 145, 146, 147, 148, 150
policy divergence, 2, 11, 14, 29, 58, 61, 65, 94, 139, 140, 146, 149, 150, 151
policy narrative, 4, 7, 9, 16, 17, 61, 64, 65, 71, 84, 89, 90, 95, 107, 109, 110, 112, 116, 130, 132, 136, 137, 138, 142, 143, 144, 145
policy paradigm, 4, 5, 51, 89, 140, 144, 149. *See also* policy regime
policy regime, 4, 5, 7, 8, 11, 15, 16, 17, 47, 48, 50, 51, 53, 57, 58, 59, 60, 61, 62, 63, 65, 66, 89, 90, 112, 116, 118, 138, 140, 145, 146, 149, 150, 152
Polish Ecological Club, 74
Ponta, Viktor, 4, 119, 126, 127, 131, 132, 136, 137
power arrangement, 4, 5, 51, 79, 82, 88, 89, 98, 100, 111, 112, 127, 136, 137, 140, 141, 144, 145, 146, 149. *See also* policy regime
producer surplus, 19

RDOŚ. *See* Regional Directorates for Environmental Protection
REACH Directive, 42
Recommendations on shale gas, 15, 41
Regional Directorates for Environmental Protection (Poland), 73, 78, 80, 81
Regional Inspectorate on Environment and Waters (Bulgaria), 95
regulation
  energy, 25, 36, 95, 150
  environmental, 3, 15, 38, 41, 123
  price, 3, 25, 38
  regulatory legacy, 16, 24, 25, 47, 71, 75, 76, 78, 79, 136, 139, 141
Repsol, 96
reserves
  shale gas, 1, 2, 11, 13, 30, 85, 93, 115
revenue sharing, 44, 73, 79, 81, 96, 133, 147
risk, 9, 41, 43, 47, 52, 109, 110, 118, 144, 152
Romania, 2, 4, 6, 7, 12, 13, 14, 17, 24, 29, 30, 40, 66, 94, 115, 116, 117, 118, 119, 120, 121, 122, 124, 125, 126, 127, 128, 129, 130, 132, 133, 134, 136, 138, 139, 140, 141, 142, 143, 145, 146, 147, 148, 150
Romania–Bulgaria Interconnector, 36
Romgaz, 115, 122, 124, 125, 127, 140
Roșia Montană, 134, 135, 138, 144
royalty, 19, 81, 96, 106, 107, 123, 130, 134, 147
Ruhrgas, 34, 37
Russia, 2, 3, 11, 13, 15, 18, 19, 22, 24, 25, 26, 27, 28, 29, 31, 34, 37, 42, 69, 72, 83, 91, 93, 100, 101, 113, 118, 119, 121, 124, 139, 153

Russian–Ukrainian gas crisis. *See* Ukrainian–Russian gas crisis

Šefčovič, Maroš, 35
Shah Deniz 2, 121
shale gas, 1, 2, 3, 4, 5, 6, 7, 8, 9, 10, 11, 12, 13, 14, 15, 16, 17, 18, 19, 20, 21, 30, 31, 33, 34, 36, 37, 38, 39, 41, 42, 43, 44, 45, 46, 47, 48, 53, 54, 56, 57, 58, 59, 61, 62, 63, 64, 65, 66, 67, 68, 69, 70, 71, 73, 74, 75, 76, 77, 78, 79, 81, 82, 83, 84, 85, 86, 87, 88, 89, 90, 93, 94, 95, 96, 97, 98, 99, 100, 101, 102, 103, 105, 106, 107, 108, 109, 111, 112, 113, 114, 115, 116, 118, 119, 122, 124, 125, 126, 127, 128, 129, 130, 131, 133, 134, 135, 136, 137, 138, 139, 140, 141, 142, 144, 145, 146, 147, 148, 149, 150, 151, 152
Shale Gas Directive, 41
Sikorski, Radoslav, 84
SLO. *See* social licence
social acceptance, 7, 45, 47, 52, 56, 66, 126, 148
social contract, 7, 66, 146, 147
Social Democratic Party (Romania), 119
social licence, 7, 12, 17, 47, 65, 66, 146, 147, 148, 149, 150, 152, 153
social licence to operate. *See* social licence
social protest, 14, 45, 87, 92, 98, 101, 105, 111, 112, 113, 114, 116, 127, 132, 133, 134, 135, 138
Southern Gas Corridor, 94, 121
Soviet Union, 2, 22, 23, 25, 26, 27, 28, 32
spot market, 19, 20, 29, 36, 37, 38, 44, 121
stakeholder inclusion, 17, 52, 53, 54, 58, 63, 80, 127, 146, 150
state-owned company, 14, 16, 24, 25, 84
Statoil, 34, 37
Sterling Resources, 115, 116, 124
Strategic Environmental Assessment Directive, 41
subsidiarity, 38
Supreme Audit Office (Poland), 76, 77, 78
Świnoujście, 69, 72

Talisman Energy, 88
taxation, 6, 73, 78, 81, 123, 130, 134
Technically recoverable resources. *See* reserves
technology
  contested, 3, 7, 12, 17, 21, 45, 53, 59, 63, 64, 65, 87, 99, 102, 103, 134, 147, 148, 150, 151, 152, 153
  leadership, 3, 23
  transfer, 12, 73, 79, 85, 151, 152
Terra Mileniul III, 124, 127, 128, 129, 133
Texas Railroad Commission, 54
Third Energy Package, 29, 97, 125
Total, 44, 67, 96
Traikov, Traicho, 93, 98, 105
Trans-Adriatic Pipeline, 36, 121
Trans-Anatolian Pipeline, 121
TransAtlantic Petroleum, 115, 124
Transbalkan pipeline, 100
Transgaz (Romania), 116, 121, 124, 125, 129, 134
transparency, 17, 32, 63, 64, 66, 89, 95, 97, 102, 107, 112, 127, 128, 136, 137, 141, 142, 145, 149
trust, 8, 17, 27, 100, 103, 107, 126, 128, 136, 137, 141, 142, 147, 148
Turkish Stream, 121
Tusk, Donald, 4, 42, 70, 82, 83

UK, 2, 6, 10, 31, 33, 34, 41, 45, 46, 71, 87, 96
Ukraine, 10, 24, 25, 26, 28, 30, 42, 91, 105, 118
Ukrainian–Russian gas crisis, 2, 23, 26, 27, 91, 105
unconventional gas. *See* shale gas
Unconventional Gas Technical Engagement Program, 3, 15, 73
Underground Resources Act (Bulgaria), 95, 96

Ungureanu, Mihai-Răzvan, 119, 127
United States, 1, 7, 18, 21, 31, 50, 56, 124
upstream, 32, 44, 59, 68, 72, 73, 76, 100, 123, 124

Vertical Corridor, 36, 121

Water Framework Directive, 41
wicked problems, 59, 62
Wintershall, 96, 124

Yamal pipeline, 28, 72

Zurawlow, 74, 87